The Social Paradox

ALSO BY WILLIAM VON HIPPEL

The Social Leap

The
Social Paradox

WHEN FINDING WHAT YOU WANT MEANS
LOSING WHAT YOU NEED

AUTONOMY, CONNECTION, AND
WHY WE NEED BOTH TO FIND HAPPINESS

William von Hippel

HARPER

An Imprint of HarperCollins*Publishers*

HarperCollins books may be purchased for educational, business, or sales promotional use. For information, please email the Special Markets Department at SPsales@harpercollins.com.

FIRST EDITION

Library of Congress Cataloging-in-Publication Data has been applied for.

ISBN 978-0-06-331925-7

24 25 26 27 28 LBC 5 4 3 2 1

For nearly four decades my students have pushed me to think harder and explain better.

This book is for you.

Contents

Introduction

Most of us know people who struggle to be happy despite having it all. I remember visiting my friend Steve after he hit it rich and being blown away by his cushy existence. As I wandered through his extraordinary home, I told him that his life was over the top. Steve admitted that it *seemed* that way but then explained to me—with a straight face—that it really wasn't. While I nibbled on his caviar and brie, I learned that the cook didn't get along with the maid, he and his wife couldn't agree on where to go on their next vacation, his daughter was waitlisted at the fancy kindergarten, and the list went on. By the time he finished I felt so sorry for him that I offered to trade places, if only to sort out this vexing cook/maid problem.

At the time I marveled at Steve's inability to see his own good fortune, but later as I was reading Frank Marlowe's wonderful book on hunter-gatherers, it occurred to me that I'm just like Steve. And so are you. Although we can't be certain, the data suggest that we're no happier than our distant ancestors, who eked out a living much like the remaining hunter-gatherers do today. The comforts, safety, and ease of our existence make us the equivalent of multimillionaires by comparison to them, but our ancestors were probably happier than we are. How could that be?

There's more than one answer to this question, but I believe an important part of the story lies in the inherent tension between our two most basic needs: autonomy and connection. As you'll see over the course of this book, these two needs play a fundamental role in nearly all aspects of our lives. Our hunter-gatherer ancestors struck a balance between these needs that worked well, but over time we've lost that balance by prioritizing autonomy over connection. This modern imbalance explains why so many people fail to appreciate the extraordinary quality of life we have created for ourselves. My hope is that the information in these twelve chapters will not only interest you but will also guide you as you make decisions about how to balance autonomy and connection in your own life.

Part I focuses on the evolution of connection and autonomy and how these needs play a central role in who we are. Part II explores major societal forces that influence and reflect these needs, including sex, culture, religion, and politics. As we'll see, these four ways of defining ourselves have a major impact on where and how people draw the line between autonomy and connection. Part III builds the case that our modern world has eroded our sources of connection by overemphasizing autonomy. These changes were brought about by urbanization, education, wealth, and science, all of which shifted the balance away from connection and toward autonomy. Meanwhile, the changing nature of marriage has at times provided a bulwark against our loss of connection and at other times exacerbated it.

Although some correction to the excessive connection and insufficient autonomy of ancestral life was probably overdue, these chapters show how the pendulum has now swung too far toward autonomy and away from connection, with important and largely unrecognized consequences for human flourishing. Part IV is designed to help you rebalance your life if you conclude that it's out of whack after reading this book. This section provides strategies for easily reconnecting in our busy, modern world and explain why rebalancing is not a one-off decision.

With this outline in mind, you're ready to start the book. But for those of you who would like a little more background, the two remaining sections of the introduction provide brief discussions of evolutionary theory and the type of evidence that serves as the basis of this book. Neither of these sections is necessary to understand the rest of the book, so if these topics don't interest you, I'd suggest you skip over them and jump straight into Chapter 1.

A Brief Note on Evolution

If you don't have a strong background in evolutionary theory, there are five points about evolution you might want to consider before diving into this book. First, when I give talks about how humans and chimpanzees evolved from the same common ancestor, people often ask why there are still any chimps. If we're a more successful species—and at the moment we obviously are, with eight billion humans and fewer than half a million chimps—why didn't they evolve into humans along with us? Similarly, when I explain how cooperation is the key to our success, people often ask why chimps aren't more cooperative. If cooperation led us to world domination, why don't our chimp cousins cooperate more?

The key fact to remember when considering such questions is that all life forms evolve to fit the niche they occupy. Chimpanzees are well adapted to life in the tropical rainforests, where being more humanlike would not make them more successful. You need only ask yourself how you'd fare if you were dropped naked into the jungle to realize the advantages of their traits for their environment.

Our ancestors' move to the savannah six million years ago forced us to become more cooperative and our increasingly cooperative nature created new opportunities to use our intelligence. Being smarter and more cooperative are both evolutionary strategies that have notable advantages but also notable costs. It takes a lot of calories to run a big brain so a big brain will only evolve when it brings in a lot more calories. Cooperation also exposes you to freeloaders—people who take more than they give—so cooperation will only evolve along with mechanisms to enforce it. If the costs of certain traits outweigh their advantages in their environment, you can rest assured those traits won't evolve. Chimpanzees wouldn't gain as much from cooperation as our savannah-living ancestors did, nor do they have the capacity to punish freeloaders that our ancestors gained when they evolved the capacity to throw stones on the savannah, and thus chimps never evolved anything close to the cooperative nature of humans.

Second, it's easy to get caught up in the here and now when evaluating evolutionary success. But what exists today is only a small slice of what has existed and many of today's species have been here for barely a blip on the evolutionary timescale. *Homo sapiens* have walked the Earth for less than three hundred thousand years. A quarter of a million years isn't bad, but some of our intermediary ancestors such as *Australopithecines* roamed Africa for ten times longer. They're not here anymore, but it's not clear if humans will be here a few million years from now either, so perhaps our comparatively small-brained ancestors will prove to be a greater evolutionary success story than we are. Maybe we're too clever for our own good.

Third, it may seem odd that our psychology evolved along with our biology, as it's hard to imagine how patterns of thought could be heritable. But ways of thinking are heritable, with most of our psychological traits showing a substantial genetic component. Even more importantly, some aspects of our psychology play such a critical role in survival and reproduction that everyone shares them—they're *species-typical*. Pretty much all humans enjoy a good meal and pretty much no humans like to eat feces or decaying carcasses. Because our bodies evolved to extract nutrition from what we call a good meal (fruits and vegetables, nuts, animal protein) but not from feces or decaying carcasses, we prefer the former over the latter.

Dung beetles and vultures enjoy foods we find disgusting because their bodies evolved to extract nutrients from such sources and to protect them from the pathogens in them. We are motivated to eat fruits and nuts and they are motivated to eat feces and roadkill. Such species-typical motives are shaped by evolution so our behaviors match our capabilities, whether we're dung beetles, vultures, or humans. Our ancestors who did not share our preference for fruit over poop were less likely to survive and reproduce, with the result that the genes leading to their unsuitable food preferences disappeared with them.

Fourth, one of the most remarkable features about evolution is that it is a mindless force. The process of evolution involves countless random genetic changes, most of which either result in no discernible effect or make matters worse. But every once in a while a genetic change makes an organism more effective in its struggle for survival or its capacity to reproduce. When that happens, the gene spreads via the enhanced reproductive success of its owner and the species evolves. Through this process, species become better adapted to their environment. In this book, I occasionally discuss evolution in terms that could be interpreted to mean that evolutionary forces have a goal or are mindfully guiding organisms toward a more successful future. Please resist the temptation to interpret evolution in such a manner. The outcome of evolution is individuals who out-compete other individuals, resulting in species that better fit their environment, but the process of evolution is entirely mindless and random.

Fifth, evolution doesn't require that organisms be aware of the importance of their survival or reproduction or motivated to survive or reproduce. Rather, animals simply need to evolve a constellation of motives that keep them safe and well fed, and that help them reproduce when the time is right. For example, fear of predators, feelings of hunger and thirst, disgust for pathogens and parasites, sexual interest in suitable mates, and feelings of nurturance to offspring will lead animals to engage in the behaviors necessary to survive and reproduce, independent of whether they have any desire to do so. We humans are the first animals to separate our sexual activity from its reproductive consequences, with the result that our species will peak in numbers later this century and then begin what may be a permanent decline. Most humans have a great deal of interest in sex, but declining birth rates around the world suggest that most humans have much less interest in reproduction. Maybe if cockroaches invented birth control they would disappear too.

The Nature of the Evidence

I make a lot of claims in this book, so you might wonder where the evidence comes from. How do we know these things? And how confident are we that these claims are true? Some of what I'll be telling you is based on a mountain of evidence and we're pretty sure we know what's going on. Other claims, however, are based on just a smattering of evidence that is glued together with a fair bit of speculation. Perhaps the most important speculation you'll see concerns the inferences I make about the behavior and psychology of our distant ancestors. Throughout this book, I'll extrapolate from what we know about the hunter-gatherers who were encountered by anthropologists and explorers in the last few hundred years to understand what our distant ancestors were probably like. Very little behavioral evidence survives from tens of thousands of years ago, so we don't know if the commonalities documented among the hunter-gatherers we've actually met are indicative of what our ancestors were like. Most of us assume that the lives of hunter-gatherers have changed very little during the time humans have existed. But we don't know that.

To give you a sense of the data that inform this book, let's consider two examples of the types of research I'll be reporting: one on ancient behaviors we can't observe directly, and one on behaviors we can study in the lab and field. To start with the ancient, one of the best examples of human connection among our ancestors is mandatory sharing. I'll discuss how mandatory sharing works in detail in Chapter 2, but at this point I want to focus on a unique form of mandatory sharing among the Ju/'hoãn (aka !Kung San) hunter-gatherers of the Kalahari Desert.

The life of a hunter-gatherer is dicey no matter where you live. Most hunts fail, meaning that hunter-gatherers are at constant risk of going hungry or even starving if they have a string of bad luck. The risk of starvation is even greater for people like the Ju/'hoãnsi, who eke out a living in an unforgiving desert environment. In response to this risk, the Ju/'hoãnsi developed a system of connections called *hxaro* that involves the maintenance of reciprocal exchange relationships over various geographic distances. Most of the partners in hxaro relationships live reasonably close to one another, which facilitates mutual aid. But one of the key features of hxaro relationships is that they almost always include several partners who live over fifty kilometers away and at least one who lives over a hundred kilometers away. Distant partnerships are difficult to maintain, but they have the huge advantage that if everyone is struggling in your neck of the woods, you can pack your bags and move far enough away so that things are likely to be different. Should it come to pass that you need to make a two-hundred-kilometer trek to escape a local famine, life is going to be a lot easier if you have an hxaro partner when you get there than if you show up hungry and friendless.

Hxaro relationships are costly in both time and goods, as you must exchange gifts to maintain the relationship and you are required to help your partners whenever they need it. But a diverse and well-chosen array of hxaro partners also serves as an insurance policy by creating a network of people you can count on that spans thousands of square kilometers. If you do your part to help your friends when they need it, you know that anywhere you're likely to go there will be someone who will help you too. Hxaro partners who are far away have the added advantage that they might be able to find you a husband or wife who is less likely to be closely related to you. The benefits of hxaro relationships, particularly distant ones, probably played a critical role in the prehistory of the Ju/'hoãnsi by increasing their chances of survival and reproduction.

For these reasons, scientists have assumed that the hxaro tradition is probably really old, long predating the visits of anthropologists who first documented it. But just because an assumption makes sense doesn't mean it's true, so we need to find ways to assess whether the system of hxaro is as ancient as it would seem. There is no way to answer this question with certainty, as there is no written evidence of hxaro from the distant past, but there are ways to investigate it.

In modern times, hxaro gifts almost always contain beadwork made of ostrich eggshells, which raises the possibility that we might be able to find evidence in the archaeological record that people gave each other ostrich-eggshell beads long ago. Such evidence would not be proof of the existence of hxaro, as perhaps people traded beads then much like we use money now. But such evidence would be consistent with the idea that hxaro networks are indeed ancient among the Ju/'hoãnsi. And "consistent with" is a good start when doing science.

As luck would have it, we can identify where an ostrich egg was laid by examining the ratio of different isotopes of strontium in the shell. Strontium is an element that animals extract from the foods they eat to build their teeth, bones, and eggshells (much like calcium). When the ratio of the different isotopes of strontium in the ostrich eggshell matches the ratio in the local rock, we know that's where the ostrich lived. As luck would further have it, archaeologists have found ostrich eggshell beads that can be dated back many thousands of years. When Brian Stewart and his colleagues analyzed the strontium isotopes in these ancient eggshell beads to identify their likely geographic origins, they found that even among their oldest samples—which were thirty-three thousand years old—the eggshell beads had traveled hundreds of miles from where the eggs were originally laid.

What do these findings mean? On the one hand, they suggest that the hxaro tradition may be at least thirty-three thousand years old. According to this possibility, the eggshell beads that traveled thirty-three thousand years ago did so for the same reasons they do today; people walked great distances to give gifts to hxaro partners and maintain a network of relationships. Coincidentally, thirty-three thousand years ago was also a time when the climate was particularly volatile in southern Africa, suggesting that the system of including distant hxaro partners might have developed in response to environmental challenges that required larger cooperative networks than were previously necessary.

On the other hand, it's also possible that these eggshell beads were used as a form of barter and traveled across broad networks as people used them in a manner akin to cash. If so, then each transaction probably involved a small geographic distance, but over numerous transactions some of the eggshell beads eventually traveled hundreds of miles. Eggshell beads would have been easy to bring along when traveling, which is important for nomadic hunter-gatherers, so people might have traded them for a variety of goods in the absence of any meaningful relationships. According to this possibility, these ancient and well-traveled eggshell beads are not evidence for the antiquity of hxaro, but rather for the fact that humans have long engaged in complex trade networks.

These data give us a sense of the challenges when we try to infer the behavior and underlying psychology of our distant ancestors. But what do evolutionary data look like when we're working with organisms that are alive in the here and now? One of my favorite examples of the ongoing interplay between evolution, psychology, and biology can be found in the life cycle of *Toxoplasma gondii*, a parasitic protozoa. *T gondii* sexually reproduce in the intestines of cats (from housecats all the way up to lions), after which their young are pooped out in the form of oocysts, which are hardy little fellows. Many of these oocysts never find another host, but if they are lucky they are unintentionally eaten by various animals, most commonly herbivores who are grazing where the cats have pooped. When that happens, *T gondii* infect the hapless herbivores by working their way into their muscles, heart, and brains.

That's all well and good for the little protozoa, but they need to find their way back into a cat to sexually reproduce. Herbivores often get eaten by cats—for example, when gnus and gazelles become lion or leopard food—but our friend *T gondii* doesn't want to leave such an important outcome to chance. *T gondii* wants the animal in which it resides to become cat chow; it does not want its host to be eaten by hyenas or vultures when it finally succumbs to old age. How does *T gondii* tilt the scales in its favor?

One of the key areas *T gondii* infects in the brains of its victims is the amygdala, which is the fear center. Rather remarkably, once it gets its little biochemical hands on this control center, *T gondii* is able to manipulate the animal's preferences. It can't cause the infected animal to run straight at a pack of lions, but it does change the animal's scent preferences; where previously the animal found the smell of cat urine to be aversive, now it finds the smell attractive. Perhaps even more remarkably, this effect appears to be highly specific. For example, when chimpanzees are infected with *T gondii* they are more likely to approach the urine of leopards—their natural predators—but not the urine of lions (who live on the savannah, so their paths don't cross) or tigers (who live on a different continent).

As you can tell from these examples, when the data are from the distant past, substantial inference is involved in their interpretation. But all data require some inference. We know from this second example that *T gondii* can manipulate its host to make the animal more likely to approach the urine of feline predators. We infer from this finding that *T gondii* evolved to influence this specific preference because such a change would make the host animal linger where cats are about, increasing its likelihood of being eaten by one. Much of the research I discuss in this book relies on similar sorts of inferences, although a lot of the data I discuss are from modern humans, allowing us to literally ask people the questions that concern us. Even then, however, there are often multiple ways their answers can be interpreted, so you'll have to decide if my interpretations are convincing.

For those of you who would like to take a deeper dive into the data than I provide here, there is a reference section for every chapter listing the research articles with the findings I discuss. Keep in mind, however, that I haven't listed the dozens (and occasionally hundreds) of additional articles that provide background and meaning to this original research. So please consider the reference section for each chapter a starting point rather than a complete list. Additionally, I don't include footnotes linking specific ideas in the text to their references, as I find such footnotes distracting when I read books like this one. Rather, I've used footnotes to highlight details and tangential ideas that might interest some readers but not others.

Part I

The Human Balance

Different species rely on different strategies to survive and thrive. Some are big enough and strong enough to go it alone (think bull elephants) while others are so weak and tiny they're only successful when they band together (think bees and ants). Like bees and ants, humans are highly effective in groups but relatively vulnerable on their own, with the result that we evolved a strong need to connect with each other. Our evolutionary trajectory put us on this path to sociality millions of years ago, but it also instilled in us a competing need for autonomy. In Part I we consider where these two needs come from as well as their impact on human happiness.

1

Competing Motives

One of the great mysteries of human psychology is why so many people struggle to be happy despite leading blessed lives. If you were to pluck a random person from almost any industrialized democracy, that person will live a much longer, healthier, and more interesting life than 99.9 percent of the people who walked this planet before us. Our distant ancestors never had so much to eat, so many choices about who would be their friends or romantic partners, so many forms of entertainment, or so many ways to earn a living. They also coped with constant danger; nearly half their children died of accidents or disease before they reached adulthood.

For most of us, a life of so much hardship and sorrow would be almost unimaginable. When I consider their existence—filled with hunger, discomfort, disease, and death—my reaction is no thanks. I'd rather give the whole thing a miss. Yet, judging from the happiness of the world's remaining hunter-gatherers (such as the Hadza of Tanzania), our ancestors apparently found a great deal of satisfaction in their lives, despite lacking everything we take for granted. For example, when Tomasz Frackowiak and his colleagues asked nearly a hundred Hadza men and women from campsites around Lake Eyasi, Tanzania, whether they were "sad," "sometimes sad and sometimes happy," or "happy" during the previous week, over 90 percent said they were happy. When the researchers asked the same question of their fellow Poles, fewer than half said they were happy.

These data raise the remarkable possibility that hunter-gatherers might be happier than we are, but perhaps their interpretation of the word *sad* is much more negative or all-encompassing than ours. For example, maybe we use the word *sad* to describe how we feel when there's nothing but leftovers in the fridge, but they only use it when they or their loved ones are seriously ill or injured. Or maybe it was a particularly good time at Lake Eyasi when the research was conducted, so the Hadza were unusually happy. Both are possible, but consider a study by Barbara de Zalduondo with the Efe hunter-gatherers of the Ituri Rainforest, which sits smack dab on the equator in the Democratic Republic of Congo. She used different measures and a longer time frame than the study by Frackowiak but found similar effects.

The Efe's cultural rules allow strong emotional displays among adults—such as laughing, crying, and sulking—making them ideal participants for a study on people's emotional displays rather than their self-reported happiness. To get a sense of their emotional life, de Zalduondo spent eight months measuring the emotional states of members of fifteen different Efe camps as they went about their daily routine, during which time she made over twelve thousand observations. Across this period, the camps experienced a significant food shortage almost 30 percent of the time. De Zalduondo wondered how a food shortage would affect the three main categories of emotional expression: (1) pleasure, (2) complaint (expressed by what we'd call whining), and (3) displeasure. Much to her surprise, there was only a small drop in pleasure from times of plenty (when they made up 32 percent of all emotional utterances) to times of hunger (when they comprised 28 percent of all emotional utterances). Perhaps even more surprising, complaints went down during times of hunger (from 36 to 30 percent), raising the possibility that people stopped sweating the small stuff when they had real concerns. Expressions of displeasure did go up substantially (from 22 to 32 percent), as people had more reason to be displeased when day after day they failed to find food. Note, however, that pleasure was expressed more frequently than displeasure during times of plenty, and only slightly less frequently than displeasure during times of hardship.

These data suggest that even when hunter-gatherers are chronically hungry, they're almost as likely to show pleasure as they are to show displeasure. Most of us are blissfully unaware of what this kind of hunger feels like, but when I think about how often my friends and I get upset over trivialities (like the time the pizza guy put barbecue sauce on my meat-lovers pizza when I specifically ordered tomato sauce!), it doesn't seem like we're any happier than they are. How is that possible?

Psychologists have devised a number of strategies to help people appreciate their blessings, many of which work reasonably well, like thanking others or expressing gratitude for one's good fortune. Activities such as these give people a short-term boost in happiness, but they leave fundamental questions unanswered. Why do the effects of such reminders fade so quickly? And why must we put in any effort at all to see blessings that should be blatantly obvious? We're like multimillionaires compared to hunter-gatherers, but we're distracted by the fact that the cook can't get along with the maid and forgetting what a blessing it is to have a cook or a maid in the first place.

The most common explanation for this strange state of affairs is that we accommodate ourselves so readily to almost any situation that our happiness depends only on *short-term* gains and losses. Whether we're hunter-gatherers or members of industrial societies, we focus on whether today is better than yesterday and what we can do to improve tomorrow, not on whether life has been good to us. According to this possibility, we're more attuned to rapid changes in our life circumstances than we are to our ongoing state of being. If that's true, we're no happier than our ancestors because all of us are caught up in the same game of comparing today to yesterday and worrying about tomorrow.

The problem with this explanation becomes apparent if we follow it to its logical conclusion. According to this possibility, a millionaire who loses a hundred thousand dollars in a bad investment should be sadder than a homeless person who finds ten dollars on the sidewalk. After all, the former has an easy life but has experienced a momentary setback while the latter has a hard life but has experienced a brief win. We don't have good data comparing millionaires when they make bad investments to homeless people when they find money, but I suspect that if the millionaire is sadder in this case, it's likely to be a short-lived effect. After consoling himself with a glass of Hennessy, the millionaire will soon feel better, and after buying himself a burger and fries, the homeless person will soon be hungry and cold again.

Thought experiments like this one clarify that our impressive ability to adapt to our life circumstances must not be the whole story. The homeless person may have plenty of moments of happiness, but life satisfaction is much higher among those who are well to do and sheltered from the elements. Given that we are so much wealthier than our ancestors, and so much better protected from the threats to their existence, it stands to reason that something in their lives must have been lost or we would all be much happier. The key question is . . . *what?* What did they have that we don't? The more I've reflected on this problem, the more I've come to believe that at its heart lie a few key details from our particular evolutionary history.

Over the last six million years, human evolution has led to a pair of competing needs inside all of us that must be balanced for us to experience lasting happiness. These needs were etched into our psychology because they supported two key goals our distant ancestors had to achieve: to *bond with others* for their mutual protection, and to *develop their own skills* to make them valuable to their group. In support of these goals, our ancestors evolved two corresponding needs that must both be satisfied for them to be happy. Millions of years later we're still driven by these needs; from childhood through to old age we have a *need for connection* and a *need for autonomy*. Unfortunately, when we pursue either of these needs, we must sacrifice the other.

Our ancestors faced this same trade-off, but their lives provided a very different balance than ours does today. As we'll see, the modern world has led to a variety of changes that have shifted our emphasis away from connection and toward autonomy. Because most of these changes emerged slowly, over the course of generations rather than years, the central role played by the tension between connection and autonomy in our modern problems has escaped the notice of almost everyone. Instead, our problems are explained with a variety of disparate theories, each of which is specific to the issue at hand and none of which attempts to tell the whole story. But almost all our problems have a common denominator: people experience a sense of emptiness when they should be fulfilled. *In short, we're often sad when we should be happy.*

This paradox is what drove me to write this book. I believe our modern world has disrupted the balance between connection and autonomy in our lives, sometimes for the better, but more often for the worse. Once we see our struggles as a product of this imbalance, our problems look different and so do their solutions. Understanding the interplay between these two needs connects the dots across seemingly unrelated problems, while simultaneously giving us a new way of looking at ourselves and our world.

The Tension Between Connection and Autonomy

As I wrote in *The Social Leap*, when local climate events forced our ancestors out of the trees six million years ago, they eventually banded together on the savannah for safety. Their increased cooperation and sociality placed us on a new evolutionary trajectory, which ensured that our most fundamental psychological need was for connection. By *connection*, I mean our desire to cooperate, form social bonds, make friendships, establish long-term romantic relationships, and attach ourselves to our group.

Our need for connection played a central role in our evolution, as it enabled us to cooperate to solve problems that we were too small, weak, or ignorant to solve on our own. Connection was a matter of life or death then and it remains critical now. Some forms of connection are new (Facebook and LinkedIn), others are as old as our species itself (having a meal with friends), but whether you're an introvert or extravert, connection is fundamental to your life satisfaction. Whenever we work together, offer or seek advice, attend a crowded party, sit side by side with a friend while studying or watching a movie, or even smile when we meet the eye of a stranger, we feel the imperative of connection. When you bask in the comfort and camaraderie of old friends, you're feeling the product of six million years of evolution.

At the same time, the need for connection we evolved on the savannah was supplemented by a need for autonomy, which remains our second most fundamental psychological need. By *autonomy* I mean self-governance; choosing a path based on your own needs, preferences, or skills; and making independent decisions. Connection makes humans effective in their struggles against predators and a harsh environment, but autonomy allows us to increase our usefulness to others.

Some species have only one way to be of value to their group or a mate (e.g., male dung beetles succeed or fail based entirely on the size of the poo ball they are able to accumulate), but we humans can adopt a wide array of strategies to make us effective group members and attractive mates. In our ancestral past, some people tried to become the best hunters or gatherers, while others strived to be the best arrow makers, storytellers, healers, or cooks.* Today the number of opportunities for self-development has expanded exponentially. Autonomy helps us create competence by motivating us to seize what we regard as the most promising opportunities for success. In short, autonomy is what makes individuals unique and efficacious.

Unfortunately, forming social bonds with others satisfies our connection needs but directly threatens our autonomy. Interdependence constrains our choices by requiring us to consider the consequences of our actions for others. In contrast, prioritizing our own goals and preferences over the needs and desires of others maximizes our autonomy but makes us unpalatable as relationship partners or coalition members.

Despite these facts, the field of psychology hasn't recognized this contradiction. Over the last forty years, psychologists Edward Deci, Richard Ryan, and their colleagues have mapped out the nature of intrinsic motivation† in their enormously important books and articles on *self-determination theory (SDT)*. This theory, more than any other, has become the foundation of our understanding of goal pursuit and life satisfaction. In this line of work, they argue that autonomy and connection are two of our most fundamental needs, but they propose that these two needs are mutually reinforcing.

---

*Keeping in mind that they still had to hunt or gather; our ancestors were all generalists.

†We are *intrinsically* motivated to do things that bring us joy in their own right (e.g., I like to rock climb, golf, and ski). We are extrinsically motivated to do things that are a means to a desired end (e.g., I consult on legal cases so I can afford green fees and lift tickets).

They base this argument, in part, on the fact that people give each other a lot of autonomy in good relationships and are often intrusive and controlling in poor relationships. As their research shows, when people are in good relationships, they're trusting and comfortable enough that they can satisfy each other's connection needs and autonomy needs at the same time, but when they're in toxic relationships they're unable to satisfy either of these needs. Although that's true, it obscures the fact that entering relationships in the first place is damaging to autonomy and pursuing autonomy is damaging to relationships.

Deci and Ryan also define autonomy somewhat differently than I do; my emphasis is on independence and self-governance whereas they focus more on feelings of volition. From their perspective, if I *decide* that I value my relationship with an art lover more than my desire to go skiing, then my decision to spend my holiday at the Louvre was made autonomously. Never mind that I'd rather visit the dentist than an art museum and am later brought to tears when I see my friends' photos from the slopes. I expressed autonomy in choosing to prioritize my relationship over my personal preferences.

At one level their approach makes perfect sense, as I was free to make my own choice (no one tied me up and paraded me through the Louvre in shackles). But this perspective ignores the fact that my decision was based on my need for connection, not my need for autonomy. Indeed, in this scenario I've sacrificed my autonomy in service of connection, so it makes no sense to say that I've made this sacrifice autonomously. Furthermore, in its emphasis on the internal feeling of exercising one's own will, their theorizing loses sight of the external factors that caused humans to evolve a need for autonomy in the first place. The need for autonomy did not evolve to give me a feeling of volition or free will. The need for autonomy evolved to motivate me to pursue opportunities that I regard as the most promising, and in so doing, find my own areas of competence.

Because self-determination theory is the dominant perspective among psychologists seeking to understand motivation, most of us have come to see autonomy and connection as mutually compatible. That viewpoint, in turn, obscures what would otherwise be obvious: *Our shift to a lifestyle that emphasizes autonomy has unwittingly but inevitably sacrificed the connections that keep our life in balance.* Autonomy without connection—the very dynamic that characterizes modern society—creates what I call *sad success stories*; people whose achievements feel hollow and unsatisfying because they don't have a tight network of friends to share them with.

The mistaken notion that autonomy and connection are mutually supporting means that problems that arise from their inherent tension are often overlooked or misunderstood. For example, it is well known that people have different attachment styles and that we can predict much about people's relationships by assessing whether they are securely or anxiously attached. Countless marriage therapists have tried to help their clients map a route to security in their attachments. But what if "attachment styles" are only partially about the capacity for connection? What if they also reflect the conflict between connection and autonomy? Rather than being a single construct, secure attachment might be better conceived as a match between both partners' connection and autonomy abilities and needs.

By way of example, two of my good friends have been happily married to each other for decades even though they barely see each other. I remember stopping by their place in the evening and finding one of them contentedly eating dinner on her own. I asked where her husband was and she said she didn't know; he hadn't told her that morning where he was going or when he was coming home. They are clearly a high autonomy/low connection couple but both of them are very secure in that understanding.

In contrast, I know other happily married couples whose lives are so enmeshed that when I run into one of them, I automatically look around for the other. It doesn't matter whether they're at the grocery or the dog park; if you need to find one it's enough to locate the other. What seems to make all these people happy in their marriages is not just the love they feel for each other, but the fact that they agree where the balance between autonomy and connection should lie. According to this possibility, there is no universal prescription for secure attachment, as the answer will vary between persons and across time and place.

To take one more example, the skyrocketing rates of depression and anxiety in adolescents and young adults have left many of us wondering why people who have more opportunities than any generation before them struggle so much with their mental health. A variety of explanations have been proposed for this problem, such as the corrosive effects of social media and helicopter parenting, and the data support a role for these factors (I'll say more about social media in Chapters 7 and 11). But we also need to consider the dramatic increases in autonomy in the last few generations and the fact that they've resulted in lost connection. The primary developmental goal of adolescence and early adulthood is to establish independence, but in the past this goal was pursued in the context of extensive and meaningful connections (that probably felt smothering at times). In the absence of preexisting deep connections, young people today follow their biological inclination toward greater autonomy without realizing that their autonomy/connection ratio is already out of whack.

The Centrality of Competence and Warmth

Autonomy and connection are critical not only to our own life satisfaction but also to how others perceive us. The two universal domains on which we evaluate others are competence and warmth. We care if people are *capable* and we care if they are *friendly*. Nearly all evaluations that matter can be subsumed under these two broad categories. As we'll see in Chapter 3, the need for autonomy evolved to facilitate the development of competence by allowing us to choose the domains in which we have the best prospects and then develop our talents in those domains. Warmth, on the other hand, is the interpersonal reflection of a person's desire and ability to connect. Warm people connect with others naturally, easily, and empathically; cold people do not.

Although competence and warmth both play a central role in our evaluations of others, we care more about warmth than competence. This is somewhat counterintuitive, particularly from an evolutionary perspective in which our ancestors were at constant risk of starvation, but warmth matters more for survival than competence (so long as people achieve an acceptable level of competence). The partner who will reliably share with you, who can be counted on to help when you need her, and who has your interests at heart is a more valuable person than the great hunter or skilled arrow maker who can't be counted on in a pinch. Consistent with these arguments, when anthropologists measure hunter-gatherers' competence and warmth, they find that people prefer to spend time with campmates who are warm and friendly over campmates who are the best hunters, even when they're heading out on a hunt together.

Competence and warmth might seem like independent qualities—after all, a person can be highly skilled or hopeless regardless of whether they are nasty or nice—but the inevitable tension between autonomy and connection ensures that it's difficult to increase one quality without jeopardizing the other. The more we pursue autonomy and competence, the more we sacrifice connection and warmth. There are only so many hours in the day, and the more time we spend practicing or otherwise honing our craft, the less time we spend connecting with others, cooperating with them, and generally attending to their needs. We pay a price in connection to develop competence.

There are exceptions to this rule, as some people develop competence so easily that they can still maintain strong relationships. But those exceptions are sufficiently rare that the more we see someone as competent, the less we see them as warm (and vice versa). For example, when people are asked to form impressions of others in experiments, they view incompetent people as warm but competent people as cold. The effect goes the other way too—learning that someone is warm makes them seem less competent and learning that someone is cold makes them seem more competent. As a result, people who are highly successful can seem cold and people who are warm and friendly can seem incompetent, even when they're not. Finding the right balance between connection and autonomy is thus critical to our own mental health as well as the impressions formed by others, and the various romantic, platonic, and pragmatic relationships we develop across our lifetime. To get a sense of where that balance should lie, let's start by considering the evolutionary origins of our need to connect.

2

Why We Connect

Before we dive into human nature, it's worth asking how evolution has shaped other creatures who face similar pressures, as their solutions to life's challenges provide a useful context for understanding our own. Many animals serve as good comparisons, and we'll consider an assortment throughout this book, but I'd like to start with a brief discussion of the kookaburra: the extraordinary bird who wakes me at dawn with its insane, beautiful laugh. Kookaburras, and birds in general, are more like humans than you might think. Similar to human babies, kookaburra chicks are difficult to raise—they are flightless and stuck in the nest—so all their meals must be hunted down and brought to them. In response to these demands of parenthood, evolution ensures that most birds form monogamous pairs in which both the mother and the father work hard to raise their highly dependent young.

Much like our human ancestors, the kookaburra is at constant risk of starvation in Australia's drought-prone countryside. Kookaburras have evolved a few interesting habits in response to the challenges of feeding their young in this unpredictable environment. First, they practice "cooperative breeding," in which earlier broods of kookaburras stick around to help mom and dad feed the latest generation. This help from older siblings increases the probability of successfully raising the chicks, as there are more kookaburras providing for the little ones. But kookaburras also add a second strategy; like many other birds, they produce an additional egg just in case.

This third egg serves as an insurance policy in two ways. First, if either of the first two eggs fails to hatch, the third egg can provide the second chick. Second, if there is plenty of food to go around, the third chick might well make it to adulthood, thereby increasing the size of the family (and the evolutionary "success" of the parents). But the third egg is definitely a backup—when times are tight, the third chick almost always dies, either of starvation or siblicide. The first two chicks are larger and if they aren't being adequately fed, their most common response is to peck their little sibling to death (using a hook on their upper beak that evolved specifically for the purpose of killing their younger brother or sister).

The proclivity to kill your little sibling when times are tough might seem like a grim solution to an unpredictable environment, but evolution works by any means necessary, with zero regard for suffering or morality. Animals that find a way to raise more offspring to adulthood pass on whatever genes enabled their success, causing the traits that facilitate survival and reproduction to dominate the gene pool whether they're nasty or nice. People are often dismayed by the violence and selfishness of their fellow humans, but it's worth keeping the kookaburra in mind when we reflect on the unflattering aspects of our own nature. Our psychology could have easily evolved to regard the smallest and weakest members of our family as disposable. Instead, we were lucky to chance upon an entirely different solution to our existential threats. By connecting with those closest to us rather than killing them, we not only staved off starvation but we rapidly became the fiercest predator on this planet. This is the story of how that happened.

Our Original Lifestyle

Once upon a time, humans had comparatively little autonomy but incredibly tight connections. For most of human history, our ancestors lived in immediate-return societies, which meant that they ate today what they killed today. Such a hand-to-mouth existence might seem precarious, but there were two good reasons for this state of affairs. First, in the hot climates in which we evolved, our ancestors had limited capacity to store meat without it going bad, so there wasn't much they could do to stockpile for the future. Second, successful hunts were sufficiently rare that potential diners always outnumbered potential dinners. Because there usually wasn't enough food to go around, efforts by hunters to store their catch would have been met with stiff resistance by their hungry neighbors.

Human societies solve the problems they face in numerous inventive ways, but the solution to this problem was always the same: mandated sharing of meat from successful hunts. This insurance policy spread the benefits of each hunt among the entire group, which substantially reduced the risk of a string of hunting failures. The rules that govern *how* the food is divvied out and by whom are numerous and often complex, but the rules never differ in *whether* it is shared with other members of the group.

For example, !Kung men trade arrows with each other prior to the hunt, with the original owner of the arrow that made the first strike responsible for dividing the meat among all the households in the camp (even if he wasn't present at the hunt). Among the Aka, the rules vary with the prey. When a bushpig is speared, the owner of the spear that struck the first blow gives the midsection to the owner of the spear that struck the second blow and the head to the owner of the spear that struck the third blow . . . unless the spear that struck the first blow was borrowed, in which case the borrower gets the rump and the owner of the spear gets the remainder. These initial recipients then divvy the meat out to their wives and their friends' wives, who share with the entire camp if the kill is large enough.

Mandated sharing of meat might seem like a triviality, or a cultural rule with only a single application, but successful hunts are one of the most important things that ever happen in hunter-gatherer communities. As a consequence, the implications of this policy are felt in nearly all aspects of life. For example, when it comes to possessions, hunter-gatherers can decide not to share an item, but only when they have very few such items. A person who owns more than a few shirts or arrows or almost anything else is often asked to share and has little choice but to agree. To decline such a request is to be perceived as stingy, which is one of the most damaging labels in the life of a hunter-gatherer. Stingy people are pariahs.

Perhaps even more importantly, when hunter-gatherers engage in the market economy with their agricultural or pastoral neighbors, they become the target of requests by friends and family to share their newfound wealth. It's human nature to dodge these requests when possible, but it's also human nature to capitulate when caught red-handed with the goods. You need only imagine how you'd respond if every time you were unloading the groceries your neighbors popped over and started taking their preferred items from your bags (and the rules of your neighborhood compelled you to share). Under such circumstances you might bring your family with you to the grocery store to eat your purchases right there in the parking lot. Or maybe you'd try to do the shopping when no one was around. Either way, you certainly wouldn't buy a few weeks of groceries and bring them into your house in the full light of day, as there's no chance you'd get to keep them. Given these social rules, hunter-gatherers consume the fruits of their labor as quickly as possible; otherwise, their hard-earned cash disappears the moment they bring it home.

Situations such as these demonstrate just how pervasive the consequences of mandated sharing are. First, mandated sharing leads people to focus on the present rather than the future, as trying to save something for later simply guarantees that you won't get it at all. In this sense, mandated sharing reinforces the cultural practice of consuming today what you kill today. Second, mandated sharing creates a highly egalitarian society, as everyone has equal rights to everything. Such societies have enormous advantages but they don't scale up in size very well, because (a) people need to know who is contributing and who is slacking so they can pick their group members wisely, and (b) decisions are all made locally and by discussion. Universal sharing works well among small groups of nomadic people who do not accumulate goods. But with the advent of food storage and agriculture, and the psychology of delayed gratification that such lifestyles require, universal sharing inevitably disappears.

People often romanticize the life of hunter-gatherers because they look out for each other so well and because they live a carefree existence that focuses on the present. We need to remember, however, that their psychology is our psychology. They do look out for each other incredibly well, but because they depend so heavily on each other, everyone is in everyone's business all the time. If a campmate isn't sufficiently generous or productive, hunter-gatherers are quick to complain. Their intense levels of interdependence lead them to monitor each other incessantly, as one weak link in the chain could threaten the entire community.

Furthermore, and somewhat surprisingly given their lifestyle, hunter-gatherers are just as materialistic as we are, a fact driven home whenever anthropologists arrive in their communities and are faced with a barrage of requests for their "extra" goods. Hunter-gatherers don't judge each other by what they own, given that no one has the right to keep a surplus for themselves (and hence there are no rich or poor hunter-gatherers), but they want things just like the rest of us. They also get tired of constant demands to share, which they occasionally dodge by hiding or immediately consuming goods they regard as special.

Tight connections are critical for human happiness because they were critical for human survival, but they are not without their costs. Because we also evolved a need for autonomy, our responsibilities to others can be an onerous burden. So let's consider why humans (and numerous other animals) are willing to pay such a steep price to connect with each other.

The Evolution of Connection

On the far northwest corner of Oahu, in a little slice of paradise, sits Ka'Ena Point State Park. You get there by driving to the end of Keawaula Beach and then walking a few miles down an old jeep trail that runs along the ocean. At the end of the trail, you encounter a surprisingly imposing fence, given that you're in the middle of nowhere, but the fence is necessary to keep Hawaii's introduced predators away, as albatrosses nest on the ground throughout the reserve. My friends and I had the good fortune to visit the park on a day the albatrosses were returning to nest, and I have never seen such happy birds in all my life.

At one nest, a mated pair was meeting up for the first time after months apart at sea. Upon sighting each other, the male and female broke into an elaborate head-bobbing dance, which was repeatedly interspersed by nibbling on each other's beaks. It was about as close to kissing as a lipless animal could get. A few minutes into their reunion dance, a neighboring female came over and joined the party. The couple seemed glad to see her, and the three of them head-bobbed and beak-clacked for so long I had to reapply my sunscreen.

Albatrosses mate for life, which, combined with their joyous reunion, raises the question of why they would ever choose to part for such long stretches of time. It's impossible to know if albatrosses get lonely, but the answer is probably not. They evolved to form long-term bonds to a partner who would commit the enormous time and energy required to raise an albatross chick, so their need for connection is strong. But the process of raising a baby albatross is so energy intensive that it leaves them weak and exhausted by the time their chick is fledged and ready to go out on its own. Depending on just how weak and exhausted they are, they often take the next season off to give their body an extra year to recover before they dive back into parenthood. To the best of our knowledge, most of this time is spent in isolation, as they soar alone thousands of miles across the ocean waves.

That's a lot of alone time, but albatrosses are highly social compared to countless other species whose only adult social interactions occur when mating. Our temptation to imbue such animals with emotions like loneliness tells us much more about what it means to be human than what it means to be them. Because the need for connection is so strong in humans, we assume that solitary animals must spend their lives sad and lonely. I remember watching a program on snow leopards as a child and feeling sorry for them when the narrator intoned that the males spend their entire adult lives alone. My pity was misguided, however, as sociality evolved in some species but not others. Snow leopards and many other animals are happiest when they're alone, as the presence of other members of their species outside the breeding season means competition rather than friendship.

The key point is that there is nothing special about connection unless it provides an evolutionary advantage. Almost none of evolution's gifts come without cost, and connection is no exception to that rule. Sociality is a disease vector, a cognitive challenge, and for many animals a relentless source of competition for space, food, or mates. When animals have little to gain by sociality, they have little reason to pay these costs. Our friend the snow leopard doesn't need help on its hunts, it doesn't need to keep an eye out for other predators, and when females are interested in mating, they yowl loudly enough to attract males from far and wide who do not need to socialize with her otherwise. As a result, snow leopards never evolved a need to connect, beyond the critical bond between mother and offspring that we see in all mammals.

Albatrosses have a strong need to connect to raise their young, but they don't need to connect when they're gliding over the waves looking for their supper. While they rebuild their strength, it's easier for them to attend to their own needs than to coordinate their activities with others. Because emotions evolved to motivate animals to do what's in their best interest, it's a safe bet that albatrosses feel a strong desire to connect when they are about to raise a chick but otherwise feel no particular need to be around each other.

Humans don't work that way. For at least the last few million years we have depended on our social networks every day of our lives for four important functions, the first two of which we share with many other social animals.

1. **Safety:** The first and simplest function of our social connections was safety. If you hang out with other members of your species, you can count on a lot more eyes, ears, and noses to be alert for predators. For many animals, such as gnus, safety is the primary goal of sociality; more gnus means more lookouts for lions, hyenas, and other carnivores who regard them as a meal. Gnus don't seem to coordinate their activities much beyond what is required for basic vigilance, but they do a great job of keeping watch at regular intervals, ensuring that someone is always on the lookout. Walking across the savannah was a dangerous pastime for our ancestors, just as it is for gnus today, so the presence of other members of their group would have greatly increased their chances of detecting predators before it was too late.

The benefit of sociality for the safety of our ancestors was more than sufficient reason for them to evolve a strong proclivity to spend time with one another. Our ancestors were at constant risk of predation, meaning that humans who preferred to strike out alone were mostly weeded out of the gene pool. Those asocial types were much more likely to end up as someone's dinner than as someone's lover, so their preference for extended alone time largely disappeared with them. For this reason, if we could clone *Australopithecus afarensis* (one of our transitional ancestors who was chimp-like in demeanor but who walked upright across the savannah three million years ago), my guess is that it would be much more social than chimps even though it didn't have the brainpower to fully exploit the benefits of sociality.

2. Effectiveness: If you ratchet up the complexity one notch, you get to the second function of sociality, which is to increase an individual's impact or effectiveness through joint activities. Animals can keep a lookout without ever conferring with one another, but animals who want to coordinate their activities for greater effectiveness need to orient toward the same opportunities and threats. In principle, this sort of goal coordination uses more brainpower than simply sticking with other members of your species, but it can be achieved without too much complexity by animals who follow a few hard-and-fast rules. For example, muskoxen form a defensive line against a single predator or a circle when they encounter multiple predators, so that the adults face outward with their large horns and the young gather behind them. This strategy works well against wolves but is a disaster when faced by humans with projectile weapons. Because the muskoxen are just following a hardwired rule, however, they can't choose a different strategy for different predators.

If the group has only a few activities that it engages in jointly, and if those activities are prompted by environmental events rather than individual decisions, they can even be enacted by animals with super tiny brains. For example, we see this form of sociality in ants and bees, who can achieve almost nothing on their own but almost anything in large groups. One ant can't carry much and one ant bite doesn't hurt much, but an army of ants can carry off almost anything and can kill almost anyone foolish enough to stay in their path.

Ants achieve their extraordinary effectiveness through incredibly large numbers, but until very recently that wasn't possible for humans. There just weren't enough of us. But what we lacked in numbers we made up for in brainpower. Humans dramatically increase the effectiveness of joint activities through planning and division of labor. By way of example, compare human hunts to those of our chimp cousins. Chimps engage in joint activities when defending their territory from other groups of chimps, when attacking other groups of chimps, and when hunting monkeys and other animals. Despite being very clever, chimps are incapable of anything but the most rudimentary division of labor. Sometimes they get lucky when hunting as a group, if one or two chimps happen to drive their prey into the waiting arms of another. But as a rule, group hunts and group warfare among chimpanzees are a mad scramble by both sides to gain an advantage, with the winner generally being the larger group.

The human capacity for planning and division of labor fundamentally altered this dynamic, making groups of human hunters or warriors far more effective than their numbers would suggest. Think of chimpanzee groups as additively effective but human groups as multiplicatively effective. If one chimp has a 10 percent chance of catching a monkey, two chimps have a 20 percent chance. But if one human has a 10 percent chance of catching a monkey, two humans have something closer to a 40 percent chance, as the second person can close off avenues of escape, distract the monkey while their partner sneaks up on it, and so on. This big-brain version of joint activities that our ancestors invented made the second function of sociality more impactful for us than for any other animal, as human groups are much more than the sum of their parts.

3. Secondhand Learning: The third function of our social connections is unique to humans because it is so cognitively demanding and because it depends so heavily on our extraordinary communicative abilities. Numerous animals learn how to survive and thrive by watching other members of their species hunt, fight, and flee, but only humans can engage in a broad category of social behaviors that fit under the umbrella of *secondhand learning*. For example, Andean puma cubs stick close to their mother until they are about two years old, during which time they tag along on as many hunts as possible. Because she has no way of explaining hunting strategies to her cubs, they must take enormous risks watching hunt after hunt as they slowly learn the tricks of the trade.

Humans engage in this sort of observational learning all the time, but we also learn through teaching and storytelling. When our ancestors wanted their children to learn how to avoid lions or sneak up on gazelles, they didn't need to bring them along to give it a try until they were confident the little ones were ready. Rather, they could start by telling stories of their own hunts and narrow escapes, secure in the knowledge that their children could learn the key lessons from their experience without having been there. No other animal can come close to achieving these goals because they're missing all three of the capacities that make them possible: (a) the ability to understand what others know and don't know, and hence what they need to tell them, (b) the vocabulary and grammar necessary to communicate complex information that transcends the here and now, and (c) the ability to imagine complex scenarios that set the scene when someone tells a story.

Secondhand learning makes an enormous difference in our lives. It might seem trivial that we can learn this way, but secondhand learning allows us to increase the body of knowledge available to our species by expanding our database with every generation. Insights gained by one member of our community at one time spread through all the members, potentially lasting for all time. If people couldn't explain things and had to show each other how to do everything, especially if they had no idea who knew what, human knowledge accumulation would essentially grind to a halt. In contrast, the first genius who figured out how to make a fire by rubbing rocks or sticks together changed humanity forever, as that knowledge was passed down through the oral history of our ancestors. Countless geniuses since then kept improving on the process until starting a fire is now achieved with the press of a button.

Secondhand learning makes human connection far more effective than that of any other animal because it exploits the enormous value of information sharing. When other animals cooperate with each other they create *positive-sum relationships* by helping when it benefits the recipient more than it costs the giver. For example, after a successful hunt, a vampire bat will regurgitate some of its food for a companion who had no luck and is at risk of starvation, but only if it has sufficient food reserves to avoid putting itself at risk. If it wasn't so lucky on the hunt, a vampire bat cannot help a hungry friend or family member even if it wants to.

Humans have fundamentally altered this positive-sum dynamic by exploiting information more than any other animal can. The secret to our success is our exceptional communication skills that allow us to share valuable information in a manner that is virtually cost-free for the provider. Unlike a vampire bat giving up its food, if I want to help you by telling you something, I don't lose that knowledge myself. Indeed, there is virtually no cost beyond a moment of my time. Because the cost is so close to zero, I can offer life-changing information to a complete stranger who will never be able to repay me ("There's a bear on the trail just ahead; you might want to turn around."). Due to the value of information and the ease of its transmission, humans have evolved to be much more connected and much more cooperative with each other than most other animals. The mere offering of such advice to a stranger in the national park will give me a warm glow—an emotional reward that evolution bestows on us when we are helpful to one another and thereby increase our value to the collective.

One of the most influential papers ever written in the social sciences is on the "strength of weak ties." The premise of the paper is that the people who are most helpful to us are often those with whom we share the loosest connections, as these loosely connected folks can provide us with information and opportunities of which we would otherwise be unaware. Although such people are not as motivated as our close friends to help us, they are much more likely than our close friends to know things we don't know. For example, when you're looking for a new job, you're often aware of the same opportunities that your close friends know about by virtue of running in the same circles. But you are unlikely to know of jobs that your more distant friends and acquaintances have run across, simply because you don't know the people they do. Consistent with this possibility, a recent experiment with millions of users on LinkedIn manipulated the "People you may know" algorithm and found that weaker ties (people with fewer mutual connections or who messaged each other less frequently) led to more job opportunities.

This finding is not only interesting in its own right, but it also reveals something fundamental about human nature and our tendency to connect. Very few animals benefit from loose ties because it is so cognitively demanding to cooperate across distant connections (dolphins manage it, but I'm unaware of any other species that benefits by cultivating and maintaining loose ties). As information machines, humans are able to help each other in ways that are so easy and low-cost that we cooperate with almost anyone with little regard for whether the favor will ever be returned. This ability allows us to create loose-knit communities who pay it forward rather than worrying about whether others might be freeloaders or will have an opportunity to reciprocate. Our freedom from such concerns leads us to initiate self-perpetuating virtuous cycles that create and enhance cooperation.

It was my first visit to an academic conference as I was beginning my career that made me realize the importance of this aspect of human helping. I arrived at my first conference not knowing a soul, but keen to meet people and embrace my new field. After the morning session we broke for lunch, at which point I went through the buffet line and headed to a table that still had an empty seat. As I sat down between two professors (whom I now know to be lovely people, but very shy), both of them were already engaged in conversation with people on the other side of me. No problem there, but what I hadn't expected was that as I pulled up a seat they would both turn their chairs away from me, leaving me sitting between two people who were giving me the cold shoulder.

It's hard to introduce yourself to someone's back, so I sat there wondering if I should eat in silence, move to another table, or perhaps tap them on the shoulder at a quiet moment to introduce myself. Just as I was weighing these (unappealing) options, an eminent professor from UC Santa Barbara named Dave plopped down at an empty table behind me, turned his chair around so that he faced me, and introduced himself. I was brand-new and a complete nobody, he could see at a glance that I was being ignored, and he kindly came to my rescue by at least pretending to be interested in me and my research.

Fast-forward twenty-five years and I'd been asked if I'd like to contribute to a celebration of Dave's career. I was delighted with the opportunity and wrote a brief story about how Dave befriended me at my first conference when nobody else did. It wasn't long before I ran into him at another conference, at which point he told me how touched he was that I wrote about that moment. He then confessed that he had no memory of it whatsoever. I have to admit, I was a bit hurt that he couldn't remember rescuing me when it stood out so sharply (and fondly) in my mind, but four years later I found the shoe on the other foot and finally grasped the underlying psychology.

A friend of mine named Brian was visiting my university to give a talk and I ran across him in the hallway. I recalled first meeting

Brian at a conference twenty-some years ago when he was a grad student and I was a young professor, so I mentioned this in passing to my colleague who was showing him around. Brian surprised me at this point by saying that it wasn't true. When I asked him if I had invented that memory out of whole cloth, he said no, but the *first* time we met was years earlier when he was an undergrad. I confessed that I couldn't recollect that meeting, at which point Brian told us how I ran into him between sessions at his first conference and started chatting with him. The people I was with suggested we should head off to lunch or some such, and I invited Brian to join us. I was clearly channeling the lesson I had learned from Dave's kindness to me at my first conference, but it was so trivially easy to include Brian in that situation that I hadn't remembered doing so at all.

Being on both ends of these experiences made me realize that because so much of the help that humans provide is effortless, the help you receive stands out much more than the help you give. Helpers forget but recipients remember. When recipients then pay it forward or reciprocate, the cycle starts again and the process continues ad infinitum. It's one of the great blessings of being human that we evolved to make cooperation so easy and hence so likely.

4. **Protection from Each Other:** There is a fourth and final function of human connection, but it differs from the other three in the sense that it protects us from each other rather than from other creatures or the elements. We are the apex predator on this planet, not because we are incredibly fierce on our own, but because we work so well together. One human is no match for a woolly mammoth, but tribes of humans ate mammoth steaks for dinner. Our habit of hunting mammoths is a clear demonstration of the second function of sociality, whereby animals increase their effectiveness through joint activities.

Of course, the capacity to achieve remarkable things through joint activity can be directed just as easily at our fellow humans.

Once humans rose to the top of the food chain, it would have taken but a moment to realize that their single greatest threat was now other groups of humans. This threat manifested itself primarily in intergroup conflict, where both archaeology and human history show incredibly high death rates from the almost continuous conflict and skirmishes between different human groups. But this threat was also relevant to existence within one's own group, where day-to-day life was an ongoing battle of wits and influence.

We need to keep in mind that our ancestors had no formal police force or governmental regulations to ensure their safety. Rather, they had to depend on themselves, their kin, and their network of friends. In this environment, a person who had numerous strong connections was much safer and more influential than a person who was on the periphery of the social group. Well-connected people could count on others to look out for their interests, take their side in conflicts, and support their plans. If disagreements in the group got really bad, well-connected people could rest assured that others would accompany them if the camp broke into subgroups and everyone struck off in different directions. Friends are important even within the relatively safe confines of one's own group; people who had plenty of good friends were much more likely to survive and thrive than people who did not.*

The need for connection that was critical to our ancestors' survival may have faded in importance, but it hasn't disappeared. Cooperative humans still achieve things that are impossible for any one human alone. There are numerous examples of this effect, but some of my favorite experiments on the importance of connection are conducted with children. In contrast to adults, young children do as they damn well please—whether they're in the lab or on the

---

* The same holds true for all smart social mammals; baboons, chimpanzees, dolphins, and killer whales all depend on their friends to thrive in their lawless worlds.

playground—so they provide a much clearer window to the soul. Children are also an excellent comparison to chimpanzees, as they tend to perform similarly to chimps on many cognitive tasks until about the age of three, at which point children leave their simian cousins in the dust.

In one of my favorite experiments on the value of human connection, small groups of chimps or children were given a three-stage puzzle box to solve. The box was designed to motivate the children and chimps by providing them with treats as they solved each stage of the puzzle. The three- and four-year-old children decimated the chimps in the puzzle-solving competition,* but the more interesting finding was *how* they did it. There were three key aspects of the children's behavior that differentiated them from the chimps: the children were far more likely to imitate the actions of other children who had figured out aspects of the puzzle, they were far more likely to teach and be taught by other children, and perhaps most interesting of all, they were far more likely to share the puzzle treats with each other. This cooperative orientation not only made the ideas of each child available to all the children, but it also maintained high levels of motivation as the children happily worked together to solve the puzzles.

The mutual support provided by children (but not chimps) to their fellow team members brings us to the final element that determines whether groups succeed: feeling comfortable enough to offer up your own ideas while remaining engaged in what your teammates have to say. When Google sifted through all the data they'd gathered on team performance among their employees, they found that the single best predictor of effectiveness was this feeling of psychological safety among team members. Teams whose members felt comfortable saying whatever was on their mind were far more effective than teams whose members worried about how

* A nice win for some small *sapiens*.

they might be evaluated if they didn't agree with each other or with their boss. It was only these psychologically safe teams that fully exploited the group mind, cooperating with each other to create something far superior to any individual accomplishment. This result—based on hundreds of teams engaging in thousands of tasks—shows that groups really shine when their members feel secure in their connections.

As was the case with our ancestors, Google engineers depend on strong connections to get the job done. But connection is not the whole story—groups also need to keep an eye out for opportunities to improve or they risk getting eclipsed by others. In the next chapter, we consider where this drive for disruption comes from.

3

Why We Need Autonomy

Humans are unique among all the animals in our capacity to envision the future. One of the most important tasks our large brain evolved to solve is to imagine what might happen later today, tomorrow, or next year and then prepare for it. Perhaps the most remarkable aspect of our preparations is when we change ourselves—when we decide what sort of person we need to be in our imagined future and then set about becoming that person. LeBron James and Stephen Curry obviously had enormous potential in basketball, but what made them two of the greatest players of all time was their recognition of their potential during childhood and their single-minded focus from that point forward to reach it. Like all truly great performers, they shaped their own future by training relentlessly to become the people they wanted to be.

Our capacity to transform our actual self into our aspirational self is a large part of the reason we evolved a need for autonomy. From early childhood, our sense of who we are becomes focused on our personal attributes that have the best chance of leading us to success. Domains in which we stand out in a positive way become central aspects of our self-definition, in part because we receive positive feedback from others. Once these abilities become central to our self-concept, they start to occupy our minds, they become more fun and interesting, and we exercise them whenever possible. They also become aspects of our future imagined self—one of our possible selves—and visions of this future self motivate us to turn our dreams into reality.

Autonomy is critical for the development of competence, as our sense of autonomy leads us to choose who we want to be, which in turn motivates us to become that person. Recall that I defined autonomy as (1) self-governance; (2) choosing a path based on your own needs, preferences, or skills; and (3) making independent decisions. The key factor linking these aspects of autonomy is the idea that autonomous decisions focus first and foremost on your own goals. Following a path that someone else lays out for you isn't autonomous unless you agree it is in fact best suited to you. Autonomy isn't about ignoring good advice, it's about placing your own preferences first.

To return to the idea of *intrinsic motivation*, we find inherent joy in developing our skills and seeing our self-improvement. That joy is highly motivating and is one of the primary tools through which evolution shapes our behavior. This effect is strongest in domains that we choose to pursue, rather than domains that are thrust upon us. For example, I might have been pleased to see that I'd become a better administrator after a few years as our department head, but I never wanted the job in the first place, so I wasn't that excited when I got better at balancing budgets and competing demands. I was excited, however, when I improved as a teacher and my students began to learn more effectively. I went into academics to teach and conduct research, not to administrate, so teaching is much more intrinsically rewarding to me than managing.

As you consider these processes, don't forget that evolution shaped our thoughts and emotions to make us successful, it didn't shape our minds to give us insight into our own motives. I believe that I find teaching rewarding because it's satisfying to help people understand new material and give them a new way to view the world. For that matter, I believe that I enjoy rock climbing because it's challenging mentally and physically to find a way up the cliff face. Had I been born with a better head for numbers and organizational charts, I suspect I'd find administration rewarding because it allows me to improve so many lives by creating a well-run department. And had I been endowed with greater height, I suspect I'd enjoy basketball more than rock climbing, and again I'd have some sort of post-hoc reason for that preference.

The key point is that evolution shapes what we find intrinsically rewarding by making us happy when we develop skills that give us our best chance of being valuable to others. Our sense of autonomy is critical in that process, as it keeps us on the lookout for opportunities to pursue domains in which we might excel. We may not realize that we're on the lookout, but when opportunity knocks, our need for autonomy ensures that we notice. I still vividly remember the first time I encountered what became my domain of positive distinctiveness. It was a single sentence uttered in 1969 that would have been entirely forgettable to anyone who overheard it. But it gained enormous prominence in my mind because it suggested a new way of looking at myself that I hadn't considered before.

I was in first grade and standing in the lunch line next to Ronny, the coolest kid in my class.* One of Ronny's friends came over to say hello, at which point (in a striking display of manners for a first grader) he introduced me, saying, "This is Billy," followed by, "He's the smartest kid in our class." It might sound strange, but it hadn't occurred to me that I might be smart or that being smart is such a valuable trait that it would be noticed and mentioned by someone like Ronny. From that day forward, I started to care about being smart. I also began paying attention to my performance in

---

* He was beach blond, a head taller than me, a superb athlete, and a really nice guy (he remained the coolest kid in school as long as I knew him—a star player on our high school football team and homecoming/prom "royalty" our senior year).

the classroom and how it compared to others, as I sought opportunities to test and develop my academic abilities.

This process also works in the opposite direction, by clarifying when we are wasting our time pursuing dreams that will never become a reality. In my own case, after practicing tennis nearly every day for years in hopes of making my high school team, I watched my little brother easily surpass me on the tennis court. It was a devastating realization when I was forced to conclude that I wasn't any good, but it allowed me to cut my losses in a sport that was never going to suit me. I remember my nine-year-old son coming to the same conclusion when we sat down over a highlights video of his team's rugby season. At the end of the video, having watched himself getting tossed around like a rag doll for game after game, he said, "Dad, I suck." I didn't want to hurt his feelings, but he did suck (particularly in comparison to one of his teammates who now plays professionally), so I suggested he take up golf.

One of the nicest things about being human is that we have multiple avenues to success: no single formula exists for a life worth living, meaning that we all have enormous value and potential. This aspect of our species is immensely positive, but we can't lose sight of the fact that our psychology evolved at a time when the ever-present risk of starvation meant we were forced to make incredibly tough decisions about each other. If someone was consuming more calories than they were bringing in, that someone was simply not a viable group member and could not be sustained indefinitely. These considerations forced our ancestors to make ruthless decisions about who was in and who was out, with the result that we're very sensitive to whether we're a net cost or benefit to our group. People who were a net plus to their group were treated well, leading them to feel a sense of warmth and belonging. People who were a net minus were treated poorly, making them uneasy and worried. Through countless iterations of this process, we evolved a strong need to be of value to our group.

Autonomy serves our need to be of value by setting us on the same path that ten-year-old LeBron took on his way to becoming a legend (and nine-year-old Jordy took on his way off the rugby field). First, autonomy allows us to evaluate our options and choose the domains in which we have the best prospects. Second, once we have chosen our domains, autonomy lets us decide how to develop the necessary skills to achieve our goals. Because most of us have numerous domains in which we could succeed, the person who is best placed to decide which domains to pursue is typically ourselves. We know better than anyone else what we enjoy, what will sustain us, and what level of practice we can put into mastering an activity. Our need for autonomy ensures that we resist the efforts of others to determine these things for us, who might otherwise unwittingly guide us in directions that are not in our best interests. If you've ever felt that what your parents or romantic partner want for you and what you want for yourself are two entirely different things, it is this function of autonomy that brought you to that realization.*

In this manner, autonomy serves as a corrective to our connection-based tendency to conform. Humans have a strong need to go along with their group—to do otherwise was to risk ostracism, which was a death sentence for our ancestors. But the enormous advantages that come with the human capacity for cumulative culture would be lost if we always conformed to group norms without considering whether there might be better ways to do things. Our need for autonomy ensures that we're always sensitive to opportunities to be the one who brings something better to our group.

That doesn't mean our need for autonomy prevents us from conforming, as most of the time we conform so mindlessly that we don't even know we're doing it. I don't make up my mind about whether to stop or go when I get to a red light just as I don't make up my mind about whether to stand facing the front in the elevator.

* Think *Dead Poet's Society* or *A Cinderella Story.*

The cultural rules to which we all mindlessly conform make life easier for everyone, so we go along with them because there's no reason not to.

In contrast, when members of our group are deciding their next activity, when they are unsure whether they should pursue option A or B, and in countless other situations, our sense of autonomy ensures that we feel entitled to pipe in with our opinion until everyone has made up their mind. But once a decision has been made, groups are no longer interested in argument and dissension, at which point we feel pressure from others if we don't conform to the group's wishes. This conformity pressure has led well-meaning people to turn a blind eye to some of humanity's most horrific actions, but conformity is necessary for groups to work effectively as a team. Particularly when groups come into conflict with one another, group members need to coordinate closely or they risk extermination. Thus, for at least the last million years, our ancestors who had a proclivity to conform were at a distinct advantage over those ancestors who were inclined to go their own way or ignore the group consensus.

Our need for autonomy is the only thing that stands in the way of this extraordinary pressure to conform, by ensuring we're ready to capitalize on opportunities to shine when doing things a little bit differently means doing them a little bit better. Not that risking our connections in such a manner is easy, mind you, just that autonomy makes it possible. I still remember my own internal struggle with these forces when I let my desire to conform take precedence over my need for autonomy. I was in third grade, learning Hebrew at our local Sunday school, when one afternoon the rabbi stopped by to check on our progress. He picked up our workbook and asked us to raise our hand if we thought the answer to the first question was A and then to raise our hand if we thought the answer was B. Mine was the only hand that went up for A, followed by the entire class raising their hands for B. The rabbi turned to me and asked if I'd like to change my mind, given that everyone else thought the answer was B. I deemed it wise to go along with my peers and said that I would. The rabbi responded to my capitulation by pointing out that I had been right the first time, that I should have followed my convictions, and that I shouldn't worry so much about what other people think.

Although the rabbi was right, we can't help worrying about what other people think because those potential ancestors who didn't care about other people's opinions were the ones who woke up alone one morning after ignoring their campmates too many times. *Alone* was a death sentence, so their lack of concern with others disappeared with them. But in third grade I was unaware of the evolutionary origins of my conformity and I burned with shame. The rabbi was a man of great importance in our Sunday school, I was a tiny nobody, and due to my cowardice I had blown the only opportunity I would ever get to impress him.

Such lessons loom large in our lives because they represent stumbles along the way as we try to find the right balance between connection and autonomy. In my own case, I vowed then and there to be more autonomous,* but I soon discovered that the lesson didn't generalize well to interactions with my friends. After a few miserable efforts to strike out on my own, I realized that connection with my friends was far more important than doing precisely what I wanted. Better to play kickball with them than baseball on my own.

Autonomy in Service of Connection

If you reflect on the logic underlying the evolution of autonomy, it becomes apparent that our need for *autonomy evolved in service of connection.* Autonomy helps us pursue a self-guided path to personal

---

* In my third-grade lexicon, I decided not to be such a "chicken shit."

development that ensured our ancestors became valuable to their group while also giving them their best chance to stand out and attract mates. Both of these ultimate goals served by autonomy are connection goals—being valued by our group and being attractive as a potential mate. This doesn't mean that our autonomy is in service of other people's needs. Far from it; we autonomously choose to do things for ourselves. But those things we choose to do make us more effective, which in turn makes us more valuable to others.

To return to LeBron and Steph, their pursuit of excellence on the basketball court made them more connected to others by making them highly valuable members of any team. Autonomy caused them to behave in ways that strengthened their connections by strengthening themselves. This might seem like a circuitous route, but evolution had almost no other way to make us a success, as most human endeavors are achieved through joint action. Our ancestors didn't survive and thrive by going it alone, so their autonomy was not oriented toward making them a strike force of one. Rather, they were successful when their skills made them valuable members of highly effective groups, so that's what autonomy evolved to help us achieve.

The irony here is that nothing is more disruptive to our connections than our need for autonomy. If we had no desire to be autonomous, we would be completely content doing whatever was in the best interest of our relationships whether that suited our personal interests or not. The partner with no need for autonomy never strays, never disagrees, is always happy to see a rom-com rather than an action flick, and the list goes on. As I discuss later in this chapter, we value autonomy in others, and hence have a tendency to see a potential partner without autonomy as spineless, but it's easy to imagine a world with no relationship problems if no one had a need for autonomy.

Still, there are many ways to pursue autonomy, some of which are more disruptive to connection than others. Most notably, humans achieve their goals by cooperating in ways that are aligned with their own needs. For example, if my partner hints at the possibility that I'm letting her down when it comes to household chores, I'm much more likely to cook dinner or weed the garden than I am to vacuum or clean up. All these tasks fit under the rubric of household chores but the former strike me as far less onerous than the latter. Similarly, if she suggests that we need a vacation, I'm particularly likely to agree if it involves skiing or the beach but I'm less enthusiastic about theater or art museums. The bottom line is that humans are more cooperative when they're asked to do things they want to do anyway—when their connection goals coincide rather than collide with their autonomy goals.

When connection and autonomy coincide, our chances for a successful long-term relationship dramatically increase. Although the field of psychology still has no idea what creates a romantic spark between two people, we've made decent progress in figuring out what makes couples successful once they get started. Sharing similar values is important. So is spending quality time together. Similar values draw people together, but quality time depends on finding things you enjoy doing together. Sometimes we discover our shared interests by learning to appreciate our partners' preferences—who knew country music could be so good?—but often we find ourselves happily partnered with people who enjoy the same things we do. We're both outdoorsy, we both love musical theater, or maybe we both enjoy long conversations over coffee and a croissant. These relationships work because we meet our connection and autonomy goals at the same time. But if I feel like I'm pulling teeth to get you to go climbing with me, or if you see my will to live evaporating during the opening scene of *Cats*, then the sacrifices we make to accommodate each other's preferences can overwhelm the otherwise fond feelings we have for each other.

Conflicts between autonomy and connection are strongest among romantic couples, but they emerge in any relationship. Friendships have an advantage over romantic couples, however, as friends can intersect only in those domains where their interests overlap. If you love to play squash, we can meet on the court once a week and have a great time every time, even if that's all we have in common. And many of our relationships are largely instrumental, in the sense that we get together because work demands it, our small children are friends, or we're next-door neighbors who want to stay in each other's good graces. Such relationships are easier, if less meaningful, because we devote less energy to them and expect less from them. But they still work better if we happen to see eye to eye.*

If we step back to survey the broad landscape of our relationships, it becomes evident that the more important the relationship is to us, the more potential there is for a conflict between autonomy and connection. And that, in turn, raises the fundamental question of how we navigate these competing goals when we don't get lucky and share the same interests. Answers to this question are numerous, which we'll see as we consider factors that cause us to weight one goal over the other, but one general answer is that humans are persuasion machines who succeed by planting their ideas in the minds of others. If I can convince you that what I want to do is what you want to do, then both of us can meet our fundamental needs together (particularly if it turns out I'm right).

But with everyone trying to nudge each other in the direction of their own goals, persuading partners and friends that your goals are really their own can be a delicate endeavor. This is why self-deception plays a role in persuasion. Nobody likes being manipulated, but self-deception allows you to achieve your manipulative goals without any blood on your hands, by acting in ways that benefit yourself while truly believing that your actions are in the ser-

---

* In the words of a T-shirt I saw on a hike, "If your friends don't run, you need new friends"—a clear weighting of autonomy in the choice of connections.

vice of others. For example, when we contribute an idea to a group project or suggest that our softball teammates adopt a new strategy in an upcoming game, we tell ourselves (and everyone else) that we're focused on the group outcome. But such suggestions are often intended to enhance our own role in collective endeavors, drawing attention to our distinctive contributions and elevating us (ever so slightly) above other members of our group.

Self-deception allows us to pursue autonomy in a manner that is minimally disruptive to connection, but the ubiquity and effectiveness of self-deception bring us back to why we evolved the need for autonomy in the first place. Whether I love you dearly or just want to exploit you, it's impossible for me to disentangle my own needs from yours. As a consequence, even the most kindhearted of advisors will push people in directions that are not in those people's best interests, simply because it's not easy to see what's truly in someone else's interest. If I value science over art, I'll encourage you to become a scientist even if you might have been a better artist. The inevitability of this sort of disagreement highlights the importance of autonomy; no one is better placed than you to decide what will make you happy and successful. Our autonomy ensures that while we may ask for advice along the way, in the end we want to be the ones who choose our own fate.

Choice as an Expression of Autonomy

Because choosing our fate is the primary purpose of autonomy, choice plays a central role in human psychology. If you want to see the impact of choice in its most unadulterated form, there's nothing like experimenting on a toddler. Until the age of two, humans are entirely dependent on their caregivers. Our ancestors typically breastfed their children until they were two, language starts to flower around then, and by two most humans have the physical coordination to get around in the world. These converging abilities and events have shaped our psychology so that our first expressions of autonomy kick in when we are about two years old. Because toddlers are not very sophisticated, they are often contrary for the sole purpose of finding their independence.

In my own children's case, it was great fun to manipulate them by playing to their need for autonomy. If I wanted them to go outside and they weren't keen, all I had to do was say that they shouldn't go out or that I really wanted them to stay in, and bang—out the door they went. If I wanted them to take a bath, I'd frame the request in terms of choice: would you rather take a bath with your ducky or your boat. Because toddlers aren't the sharpest, it never occurred to my kids (at least not until later) that they didn't have to choose between a duck and a boat but could instead opt for *no bath at all*.

Toddlers may wear their hearts on their sleeves, but the autonomy they express in these simple situations mirrors the need for choice that the rest of us experience throughout our lives. When people demand something of you—removing all semblance of choice—it's an automatic response to fight back via a process known as *reactance*. Even if you'd been inclined to do what is now demanded of you, the mere act of someone trying to remove your choice causes you to no longer want to engage in the behavior. People not only fail to comply in these instances, they also are much more likely to undermine those who curtail their freedom.

The same thing holds when people tell you what you cannot do. The idea that nothing is so tempting as forbidden fruit is as old as the Garden of Eden—almost everyone struggles to resist doing what they're not allowed to do. There are many wonderful experiments that show these effects, but my favorite example is encapsulated in this photograph I took outside my hometown of Anchorage, Alaska. A lot of road signs in Alaska are shot up, but this one was more bullet-riddled than most.

Pushing back when someone restricts our freedom reflects one aspect of our need for autonomy, but on the more constructive side,

Figure 3.1
Reactance makes forbidden targets more tempting—this "no shooting" sign is almost illegible from all the bullet holes.

we benefit when we have choice or control over our world. Countless experiments have shown the value of choice and control, but the classic is Ellen Langer's work in the mid-seventies on lotteries. In one of her most famous experiments, she asked people if they wanted to buy a lottery ticket for $1 with the possibility of winning $50. Once people agreed to buy a ticket, she let them pick their ticket or she picked it for them. Later, on the morning of the lottery, participants were approached again and told that someone else in their office was keen to get a ticket but they were sold out. Participants were asked if they would be willing to sell their ticket to this person, and if so, to name their price. When people had not chosen their ticket, the average price they requested was $1.96 (suggesting they understood the law of supply and demand). In contrast, when people had chosen their own ticket, the average price they demanded was $8.67. This experiment, and thousands that followed, provided clear evidence that people value things they choose for themselves more than things others choose for them. In this experiment the choice was clearly meaningless, but choice is important to people because it typically leads to better fit and greater value for the chooser.

Privacy and Autonomy

Like choice, the need for privacy is fundamental. We may occasionally wish we could read the minds of others, but we all have thoughts we're only too happy to keep to ourselves. The same holds for many of our behaviors. Most of us are inclined to close the bathroom door when we go in and most of us would prefer that our intimate moments with our partners are not webcast to the world. These preferences feel so ingrained that the need for privacy in such instances would seem to be a human universal that has been stable over time and place. The data suggest otherwise.

First, we know that our distant ancestors lived in caves and other semi-public spaces that would have made it difficult if not impossible for people to get away from their campmates and children when they wanted to have an amorous moment. Thus, it's a virtual guarantee that friends and family would have been a frequent (if often disinterested) audience to our ancestors' fornication. The same holds for defecation. With the invention of the flush toilet, we started peeing and pooping in purpose-built rooms, but that's a very recent state of affairs. As Steven Pinker points out in *The Better Angels of Our Nature*, books of manners from medieval Europe often concerned themselves with such bodily functions, advising people not to "relieve yourself in front of ladies, or before doors or windows of court chambers," nor to "greet someone while they are urinating or defecating." Such seemingly unnecessary advice suggests that as recently as a few hundred years ago, our ancestors were unfazed by the public nature of their toileting behaviors.

Nothing feels more basic than the need for privacy while we have sex or use the toilet, but these examples from our past suggest that nothing could be farther from the truth. So what is privacy and why do we feel a need for it? The answer to these questions can be found in our capacity to lie, which is our ability to intentionally deceive someone about the state of the world. Like the need for privacy itself, the capacity to lie is not inborn, but rather emerges around the age of three or four when children develop theory of mind (which is the realization that the contents of other people's minds differ from your own). As adults we intuitively know that not everyone shares our thoughts and preferences, but children need to learn this fact. Once they learn it, they immediately realize that the discrepancy between what they know and what others know can be exploited by lying, thereby allowing them to get out of trouble or gain unearned benefits.

The most fundamental form of privacy, which all humans experience, emerges from the realization that because thoughts aren't universally shared, we can choose which information to express and which to hide. If Billy has a crush on the cute girl who sits opposite him in his eighth-grade science class, he might want to tell his friends about it. If he's attracted to the sheep on the farm where he works during the summer, he might want to keep that to himself. This difference in the proclivity to withhold or share such information is driven by the consequences of that information for his reputation. Billy's friends are likely to be sympathetic to his crush on Michelle, but aghast at his interest in Fluffy the sheep. The former information might even increase the centrality of his position in his middle school social network, but the latter would assuredly lead others to devalue him, causing him to feel ashamed.

This logic shows us that the need for privacy is fundamental after all, but the only thing that's fundamental about it is our desire to shield our thoughts and actions from others when we worry they would cause us harm. Because social norms differ so dramatically over time and place, the thoughts and behaviors that we keep private also change over time and place. When I was in high school and college in the seventies and eighties, there was nothing more damaging to your reputation than being gay. I had a few friends who were gay, but I had no idea because they weren't telling anyone. Many of them even denied it to themselves in their desire to avoid almost universal societal derision. Fast forward to the 2010s and a student in one of my large lecture classes raised his hand to ask a question, in the course of which he outed himself as gay just because it provided a humorous way to phrase his question. Such changes in what people share with others show us that the need for privacy may be universal and fundamental, but the actual topics we keep private are not.

Once we understand that the need for privacy emerges in large part from our desire to maintain our reputation, we also realize that privacy can be in service of our autonomy. Privacy of thought is fundamental to many aspects of human functioning and is not necessarily tied to autonomy. For example, I may want to avoid boring you with my extensive ruminations on my stamp collection, or perhaps I don't want to waste your time with my half-baked plans for the holiday until they're fully baked. But privacy of action depends entirely on the need for autonomy. If it weren't important to choose your own destiny, or even just to do whatever you wish on a Friday afternoon, humans would only need privacy of thought. Under such circumstances, the only protection we would need from others is the assurance that they are unaware of our socially inappropriate proclivities or ideas. Privacy of action would be immaterial, because our actions would never be directed by our socially inappropriate thoughts. But in a world where people want to behave in ways that at least some of their peers would regard as unacceptable, privacy of action is one of the only ways to maintain autonomy in the face of disapproval. Indeed, privacy of action is often necessary to safely enact our autonomous goals when others disapprove.

These ideas are evident in the justification that officials typically provide when they protect our privacy (if they choose to do so). For example, here in Australia the government defines privacy as "a fundamental human right that underpins freedom of association, thought and expression, as well as freedom from discrimination." This definition goes on to note that "privacy includes the right: to be free from interference and intrusion, to associate freely with whom you want, to be able to control who can see or use information about you." Concerns about privacy center on the same issues in the United States, as is evident in the first legal defense of privacy put forward in an 1890 *Harvard Law Review* article by Samuel Warren and future Supreme Court justice Louis Brandeis. Warren and Brandeis's primary concern was for the freedom of consenting adults to behave as they chose—within the bounds of the law—without worry of reputational damage. When we reflect on why democratic governments feel compelled to safeguard people's privacy, it becomes clear that the essence of privacy is the protection of autonomy. Privacy allows you to do what you want to do while preventing others from knowing about it.

Autonomy, Egalitarianism, and Inequality

Given our common ancestry, it's no surprise that we share a number of traits with our chimp cousins. After all, we also share almost 99 percent of our DNA with them.* But at the same time, it's dead obvious that evolution has shaped us differently; no one would confuse a chimp with a person. What's not so obvious is which of our

---

* Although that's a remarkable degree of overlap, it's worth remembering that about 60 percent of our genome has a counterpart in the banana, in which the proteins encoded by their genes and ours are 40 percent identical. Evolution is conservative— typically repurposing rather than starting from scratch—hence there's bound to be a lot of genetic overlap between all living things.

psychological differences from chimps are a matter of culture and which are a matter of biology. The fact that we're way smarter is easy to understand; that psychological difference can be attributed to biology in the form of our much larger brains. But what about our tendency to try to put ourselves above others? On the one hand, the fact that societies vary a great deal in how hierarchical they are raises the possibility that the tendency might be cultural. On the other hand, the fact that people in all societies try to exert authority over each other, combined with the fact that chimpanzees are highly hierarchical, raises the possibility that the propensity toward hierarchy might be written in our genes.

This question becomes critical as we seek to understand the evolution of autonomy. Are humans like the alpha chimp, who rules his group largely by manipulation and intimidation and is always striving to put himself above others? Or are humans more egalitarian and inclined to see each other as equals? The answers to these questions are *yes* and *yes*: we have strong hierarchical propensities that are reflected in our desire to put ourselves above others, but we also have strong egalitarian propensities that can be seen when others try to put themselves above us.* Both of these tendencies are rooted in our conflicting needs for autonomy and connection. To get a sense of how these processes work, let's take a brief detour through what I'll refer to as the *four stages of human history*.

The first stage began with the emergence of anatomically modern humans about 250,000 years ago and continued for over 200,000 years (and continues today in a few isolated pockets of humanity). This is the stage I've been discussing whenever I refer to immediate return hunter-gatherers, who eat today what they kill today. As described in Chapter 2, the mandated sharing that is universal among hunter-gatherers serves as a powerful leveling device. All humans in such communities have essentially equal rights once

* Hypocrisy is baked into our psychology.

they reach adulthood, and no one can force anyone else to abide by their preferences. If someone gets too bossy, everyone else can simply choose to ignore them or abandon them and join another group. Indeed, bossiness and an inflated sense of self-importance are some of the biggest concerns among hunter-gatherers, with many of their cultural rules aimed at squelching such tendencies. For example, the best hunters are often the most modest about their success, as they're acutely aware that people are continuously monitoring their behavior to make sure they don't act like they're more important than others.

One of the clearest reasons for our success as a species is that our behavioral repertoire is so flexible that we can adapt ourselves to almost any environment on Earth, and indeed humans occupy more varied environments than any other species. Thus, it's no surprise that hunter-gatherers changed their lifestyle as they started to occupy new parts of the globe, particularly as they moved away from the tropics to cooler environments where food could be difficult to secure in winter. As our ancestors began occupying such environments, they shifted to a world in which stockpiling was necessary to survive the leaner times that were common in winter. That shift, in turn, required a change in our psychology to accept the notion that surpluses were no longer public goods that should be consumed immediately, but rather a necessary foraging goal of all families. Thus began the second stage of human history.

Because surpluses attract others who see an opportunity to take by force or subterfuge what they haven't gained through their own labor,* stockpiling required people to start organizing themselves in their mutual defense. Such organization almost inevitably

---

* We see the same behavior throughout the animal kingdom. For example, ants are one of the most successful species at stockpiling food, with the consequence that an enormous number of species evolved to make a living solely by exploiting ants' resources. Wherever we find food storage, we find someone trying to take it.

resulted in hierarchy, as some individuals had the good sense, good fortune, or simply the necessary ruthlessness to stake a claim to valuable locations, to which they then gave others access in return for their labor or protection. Among hunter-gatherers who have the capacity to stockpile resources, such as in the Pacific Northwest where large salmon runs were common, we see archaeological signs of hierarchy, such as bodies interred in graves with or without status markers or prestige goods like jewelry.

The third stage of human history followed naturally from the second with the advent of agriculture (and even greater capacity for food storage) about ten thousand years ago. The demands and opportunities of agriculture soon led to the emergence of cities, fiefdoms, kingships, etc., as vast communities grew around the best farming areas (such as the Nile River delta). In this third stage, the inequality that many people had become accustomed to in Stage 2 became a dominant feature of life. That inequality, in turn, demanded an entirely new psychology to justify the enormous discrepancies in living conditions. For example, pharaohs were regarded as gods rather than humans despite the fact that they lived and died, crapped and copulated, just like everyone else. As egalitarianism all but disappeared in such environments, people began to regard other humans as valuable or worthless as a function of their resources.*

Not only was 95 percent of the world living in abject poverty during this time, but the idea of universal human rights had completely disappeared. Peasant farmers lived at the whim of their landlords who could and did treat them like property. If a crime was committed, judicial torture was standard practice for extracting a confession, suggesting that people were regarded as guilty until proven innocent (and that torturing someone for a crime they

---

* Steven Pinker's *The Better Angels of Our Nature* does a superb job laying out just how bad the average person's life became during this third stage of human history.

might not have committed was not a big deal). Perhaps even more telling, the courts made no distinction between property and people. For example, linguistic analyses of court records from London's Old Bailey showed no distinction in prosecutors' arguments regarding property crimes versus violent crimes until the late 1700s. During this time period, more than half of the judicial executions in the American colonies and states were for property crimes. Lastly, when people were executed by the state, it was often a public spectacle that could involve the most ghastly torture, carried out in large part for the amusement and edification of the audience that would gather for such events.

All this changed during the Enlightenment, at the onset of the fourth stage of human history, when (most) people rediscovered the intrinsic value of their fellow human beings and the universal rights that hunter-gatherers took for granted. This dramatic change in how we perceived other humans—from fiercely egalitarian to property and most of the way back again—raises the question of how it could have happened. How is it that hunter-gatherers were able to stymie the hierarchical tendencies we inherited from chimpanzees, despite the complete absence of government or official law enforcement? And how did we lose that capacity during the first few thousand years of government and official laws? Perhaps most importantly, how did we regain it? The long answer to these questions can be found in the complete history of our species, laid out in a number of superb and ambitious books.*

But the short answer to these questions can be found at the intersection of our needs for autonomy and connection. If humans

---

* There are too many such books to provide a thorough list here, but a few of my favorites are: *Guns, Germs, and Steel* (Jared Diamond); *Hierarchy in the Forest* (Christopher Boehm); *The Wealth and Poverty of Nations* (David Landes); *The Weirdest People in the World* (Joseph Henrich); *War Before Civilization* (Lawrence Keeley); *The Better Angels of Our Nature* (Steven Pinker); and *Why Nations Fail* (Daron Acemoglu and James Robinson).

had a need for autonomy but not connection, we would have no need to control anyone else because perfect autonomy could be achieved on our own. Snow leopards have no need to control each other, they simply avoid each other unless they're fighting over a scarce resource. But when the need for autonomy coincides with a strong need for connection, perfect autonomy can only be achieved by controlling others. If you want to pursue your own interests but you also want company, you need to persuade or demand others to join you whenever your interests don't overlap perfectly. Only when others conform to your preferences can you meet all your autonomy needs while maintaining your connections. And that is only possible if your friends are perfect soulmates or you're in charge.

When you consider this clash between autonomy and connection, you realize the impossibility of achieving both. Almost all humans have the desire to run their own lives, but our interdependence means that this desire for autonomy inevitably manifests in a desire to control others as well. We may couch the desire to control others in benevolent terms—I'm just looking out for you, this dinner/movie/vacation will be more fun, etc.—but the stories we tell ourselves don't change the fact that what we're really seeking is control so we can satisfy our own needs. Of course, it's mathematically impossible for everyone to control both self and other, and thus life becomes a contest between those seeking control over others and those fighting against it.

The irony of these simultaneous struggles may be lost on us, but whenever we engage with others, our underlying goals—stripped to their most basic form—are to (1) control them, so they'll accompany us on our chosen path; and (2) control ourselves, which means we're trying to stop others from controlling us in their efforts to get us to accompany them on *their* chosen path. The consequence of these pervasive but competing goals is that whenever local conditions favor the accumulation of power in just a few hands, hierarchy emerges quickly and fiercely. Stockpiling was the initial ecolog-

ical requirement that led to hierarchy, with subsequent population growth and the resultant cultural abandonment of consensual
decision-making cementing hierarchies that had already developed.

But hierarchy will always result in more people at the bottom
than the top, thereby stymieing the autonomy needs of more people
than it satisfies. And that, in turn, guarantees there will always be
enormous counterpressure against hierarchy. We may never again
achieve the egalitarianism enjoyed by our hunter-gatherer ancestors, but our need for autonomy will always push us toward universal human rights and away from dictatorship, even when combined
with a need for connection that leads us to seek control over others.

When we consider the evolution of autonomy and connection, as
well as the tension between them, it's easy to imagine that we're all
locked in an identical struggle between these two forces. Although
there's some truth to that possibility, these two basic motives are
also sensitive to other aspects of our lives, as they are reflected in
and influenced by major aspects of our identity. Whether autonomy or connection emerges as the winner of that struggle varies
with different aspects of the self. We turn now to a consideration of
how identity shapes and is shaped by our connection and autonomy
needs.

Part II

The Major Forces Shaping Autonomy and Connection

Part II examines sex/gender, culture, religion, and politics and their link with autonomy and connection. These four categories define us, whether we like it or not, as they play a major role in who we are and how others treat us. These four categories also influence and reflect our own weighting of autonomy and connection. As you'll see, some of the most important divisions in our world can be boiled down to the balance we strike between autonomy and connection.

4

Autonomous Men, Connected Women

Sex differences occupy a unique position in human psychology. Our biological sex at birth (from this point forward, shortened to just the term *sex*) is one of the most readily noticed features about us, yet it is wildly controversial to assume that differences between individual males and females are in any way a product of our biology. Before we discuss sex differences in autonomy and connection, it's worth considering the sources of this controversy, as it will inform our understanding of what it means for men to be more autonomous and women to be more connected.

From my perspective as a research psychologist, I see two major reasons why people get confused and upset when we discuss sex differences. First and foremost, any attempt to categorize a species as variable as humans into only two types is doomed to failure. As we'll see throughout this book, even self-chosen binary categories are incredibly messy. For example, to be religious means different things to different people, and many self-defined atheists regard themselves as highly spiritual. Similarly, people on the political left or right often disagree with each other as much as they disagree with people on the opposite side of the aisle. Indeed, some people whose attitudes place them on either end of the political spectrum switch at some point to be equally extreme on the opposite end, without ever occupying the space in the middle. In short, humans are much too complicated to be easily categorized as an X or a Y.

In the case of males and females, the enormous variability *within* each sex tends to swamp the variability *between* the sexes. In statistical terms, we can think of this problem as one of overlapping distributions. The averages of men and women may differ a fair bit, but wherever we find a lot of men who have a particular trait, we're likely to find at least some women and vice versa. For example, if we consider the largest human sex differences that exist—such as in height or physical strength—a moment's reflection reveals that we know plenty of women who are stronger than plenty of men, and plenty of women who are taller, too. This means that even as science identifies numerous domains in which men and women differ, it simultaneously documents the fact that those differences only rarely describe most men and women, much less all of them.

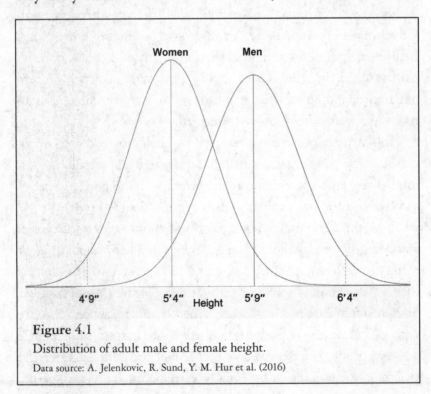

Figure 4.1
Distribution of adult male and female height.
Data source: A. Jelenkovic, R. Sund, Y. M. Hur et al. (2016)

To get a sense of what that really means, let's take a closer look at human height.* In Figure 4.1 we see the worldwide distribution of male and female height. This figure shows that the average difference between men and women is five inches. That's a pretty big difference, which is why no one who has spent any time on our planet would be surprised to hear that men are taller than women. But let's compare that to the variability within the sexes. Ninety-five percent of women are between four-nine and five-nine, meaning that even when we exclude the tallest and the shortest people, women on the tall end of the spectrum are a foot taller than women on the short end. The same holds true for men, who range from five-four to six-four when we consider the 95 percent of men in the middle. The key here is that the variability within men and women—even when we exclude the people on the ends of the distribution—is more than twice the variability between men and women. And let's not forget that height is one of the most sexually dimorphic traits (that is, it shows one of the largest differences between men and women). Most sex differences are much smaller than that.

The fact that there is greater variability within the sexes than between them virtually guarantees that judging people by their sex will lead to inaccuracy, because the chances are strong that a person may not be a good match to the average of their sex on any one trait. Even if I'm a prototypical male in twenty-seven different domains, that doesn't mean I'll be a good match on the twenty-eighth. It's also a fundamental moral tenet that we shouldn't judge individuals based on the average properties of their group. If I have

---

* A topic near and dear to my heart, as I was always the shortest kid in my class. I started my freshman year in high school at four-foot-eleven, shorter than every girl in my class but one. I was thrilled when I grew to a towering five-six by the end of my sophomore year, as I was now a smidgen taller than the average female and that struck me as sufficient (although I still dream about being six-one someday).

a job for which I need to hire someone who can carry a heavy load, it would be immoral (and foolish) to rule out all the female applicants without testing whether they can carry that load.

The substantial variability within the sexes ensures that squeezing the whole world into just two identities will leave many people feeling they're not a good fit. To take another example, sex differences in facial appearance are large, as the bone structure of the face is influenced by testosterone and estrogen during puberty.* In this case, *large* means that we can identify the sex of unknown men and women with 95 percent accuracy in photos if we erase everything in the photograph except the face. Such a high level of accuracy probably won't surprise you, but an error rate of 5 percent is still a lot of people who do not look like their biological sex (400 million humans to be precise—more than the population of the entire United States). If 400 million people don't *look* like a good match to their biological sex, it's safe to assume that lots of people often don't *feel* like a good match to their sex. That doesn't mean that billions of people wish they belonged to the other sex, but it does mean that millions or even billions of people probably don't feel like they match the long list of assumptions of what they should be like based on their sex.

The second source of controversy regarding sex differences in psychological traits lies in the fact that we don't know with certainty where they come from. Are they a product of sex differences in biology or are they an artifact of culture and socialization? It's difficult to answer this question definitively because ethical principles preclude us from doing the kind of experimental work on humans that would quantify the impact of nature and nurture. We can't randomly assign children to be exposed to male or female hormones at different points in development. Nor can we randomly

---

* Adult male testosterone levels are nearly twenty times higher than female testosterone levels and sex differences in estrogens are of a similar magnitude in the opposite direction.

assign them to be raised by their parents as a typical male or female, to be treated by their peers or teachers as a typical male or female, and the list goes on. If we could do all those things, we'd be able to measure their impact and see what really leads to the kinds of sex differences that interest us.

But that doesn't mean there's nothing we can do. After all, we can't randomly assign people to become smokers or alcoholics, but we still know that cigarettes and excessive alcohol consumption are bad for you. So, let's turn briefly to the origins of sex differences in autonomy and connection to see what evolutionary theory has to offer before we dive into the consequences of these sex differences for how we work and play.

The Evolution of Sex Differences in Autonomy and Connection

The roots of the controversy surrounding sex differences provide a critical backdrop to our discussion of autonomy and connection, as they highlight that psychological sex differences come from multiple sources and do not come close to describing all men and women. But sex differences in autonomy and connection do describe the averages of the two groups, and like height, these average differences are large enough to ensure that people notice them and act on them. Indeed, the differential weighting of connection versus autonomy is one of the largest psychological differences between the sexes, so it probably won't boggle your mind to learn that women tend to be more connected and men tend to be more autonomous (from here forward I'm going to drop the "tend to," as I'll be discussing average differences between the sexes, not how any one woman or man feels). It probably also won't surprise you that both biology and culture/socialization play major roles in the creation and maintenance of these sex differences, but here I'm going to focus on the role of biology.

Most of the recurrent problems that men and women have faced throughout our evolutionary history are pretty much the same regardless of sex: how to stay warm or keep cool; how to avoid predators and pathogens; how to get enough nutritious food to eat; how to avoid accidental injury; how to find your way home; how to stay in the good graces of your group so they don't give you the heave-ho, etc. Each one of these recurrent problems has left a mark on our psychology as we have evolved species-typical ways of solving them (such as fear of predators to avoid being eaten; fear of heights to avoid falling to our death; fear of rejection to avoid getting the heave-ho). Men and women tend not to differ much, if at all, in such domains.

But some of the recurrent problems that we've faced are only relevant to one of the two sexes, due to differences in our reproductive biology. For example, because insemination is internal, men never know for sure if they're the father, but women always know they're the mother. Salmon don't suffer the paternal uncertainty felt by human males, as the male salmon spray the eggs with semen as they are laid and witness insemination in real time.* As discussed in Chapters 2 and 3, all humans have a need to connect and all humans have a need for autonomy. These needs are species-typical, like fear of predators, heights, and rejection. But women have extra reasons to connect and men have extra reasons to be autonomous, with the result that on top of the universality of these needs, women have more of one and men have more of the other.

What are women's extra reasons to connect? They largely surround childcare. Because women put so much effort into reproduction (nine months of gestation and two years of lactation in ancestral societies just to produce a toddler who still needs an enormous amount of parental care), they typically need lots of help raising their children. Men provide much of the help in the form

---

* Not that this paternal certainty makes any difference, as they promptly drop dead.

of hunting and protection, but the fact that men head out to hunt means that someone else needs to help with the children while they're gone. That someone else will have to be a woman. As a consequence, those women who form strong connections to other women are more likely to benefit from what scientists call alloparenting, which is when people other than the biological parents help raise the children (our ancestral form of babysitting).

The capacity to attract high-quality alloparenting might not seem like a big deal, but approximately 40 percent of our ancestors' babies didn't survive childhood. Any factor that increased the quality of care our ancestors' children received would have improved those odds. Women who felt a strong connection to other women would have been ideal providers and recipients of alloparenting, enhancing the survival of each other's offspring. Because the male role in child-rearing was typically much more indirect, via the provision of food and protection, men did not have the same evolutionary pressure on them to form tight connections to a small group of others who would provide them with alloparenting.

Given the reciprocal relationship between connection and autonomy, the fact that women are more connected than men suggests that men will be more autonomous than women. But there is an additional evolutionary reason for this sex difference as well. As discussed in Chapter 3, autonomy allows us to increase our usefulness to others and be of value to a potential mate. Both sexes must mate to produce offspring, but because females put much more effort into reproduction, males typically compete for female attention and interest. This effect holds across the animal kingdom, with the greatest male-male competition occurring in mating systems in which males devote their energies to mating rather than parenting (such as elephant seals and gorillas). Human males typically parent more and have fewer simultaneous partners than gorilla males, but men still compete for women more than women compete for men. Indeed, when we look at our male-inherited and female-inherited

DNA, we see far more female ancestors than male ancestors. These data suggest that many males failed in these mating competitions while other males were highly successful, siring children with multiple females. Some men were silverbacks but many more were monks.*

These genetic data show us that the evolutionary pressure on males to attract and retain a mate is stronger than it is on females. And that, in turn, means that autonomy will be of greater importance to men than it is to women, due to the critical role autonomy plays in making us desirable. By enhancing our competence, autonomy gives us an edge in mate competition. Given the greater evolutionary importance of autonomy for men and connection for women, it's no surprise that parental and peer socialization tend to reinforce typical sex differences in these traits. As a result, these sex differences are large and culturally universal. Because these sex differences influence how men and women interact with each other in numerous ways, we'll examine their impact in a few of the major domains of human experience.

Sex and Friendship

Girls and women have tight, small groups of friends. Boys and men have loose, large groups of friends. That state of affairs (which also describes the sex differences we see in many of our primate cousins) might seem trivial but it has far-reaching consequences. First and foremost, tighter networks expect greater loyalty from each other and are less forgiving of betrayal. We expect our closest friends to have our back and are surprised and hurt when they don't. Imagine how you'd feel if you discovered your spouse is having an affair with your best friend versus your occasional tennis partner. In the

---

* We see the same thing on Tinder, by the way, where a minority of men get the majority of interest but the interest in women is spread much more evenly.

former case, you'd not only be angry at your spouse but you'd also feel deeply betrayed by your friend. In the latter case, you'd probably look for a new tennis partner, but your anger would be reserved primarily for your spouse.

Because women have smaller circles of closer friends than men do, they expect more from their friends. Their friends are meant to take their side in any conflicts that emerge and are not meant to oppose them in any meaningful way. In contrast, because men maintain large circles of relatively loose friendships, they expect much less from their friends. Loosely connected people often find themselves in disagreement or on opposite sides of a conflict simply because they have different priorities and loyalties, so men are unsurprised and unhurt when they don't see eye to eye with all but their closest friends.

When our bonds with someone are not very tight, we're more responsive to the situation we're in than the person we're with when choosing a course of action. As a result, disagreements and conflicts among boys and men typically center on the situation rather than the person, which means that boys and men get along just fine after they've been in conflict with each other. New situations create new alliances. For example, a college friend of mine had a business partner who discovered it would be more profitable to unravel their relationship than to close the deal they had agreed on. Less than a year later, that ex-partner encountered another opportunity that required my friend's expertise. Despite having let him down previously, the ex-partner had no problem going back to my friend to pitch the new opportunity. And despite my friend having been shortchanged the last time, he had no problem joining the new business venture (with a few extra contractual safeguards). I'm not saying they became bosom buddies, because they didn't, but both parties were driven to form a pragmatic alliance by the opportunities of the moment, with little thought of their larger relationship. Girls and women sometimes make the same decisions, but they are less forgiving of each other and typically need to repair their relationship after such a betrayal if they are to continue working together.

These sex differences in friendship networks are related to and compounded by sex differences in autonomy and connection. Women are more connected than men, so their relationships are more important to them and weighted more heavily in their life decisions. When you add in the fact that women have a smaller network of closer friends, it becomes apparent that female friendships demand more sacrifice than male friendships. If two female friends get together every week over lunch to catch up or exercise, it would be a notable slight if one of them canceled at the last minute because a better opportunity emerged. We don't expect that sort of behavior from our close friends, as they should prioritize getting together with us over the activity itself.

In contrast, men are more autonomous than women, so their own preferences often outweigh the demands of their connections. When you add in the fact that men have a larger network of looser friends than women, it becomes apparent that male friendships can easily accommodate men's greater inclination to pursue their independent interests. If my lunchtime squash partner cancels at the last minute because he got invited to play basketball with someone else, I'm relatively unfazed by his prioritizing the sport over our relationship because I feel the same. I, too, would grab a better opportunity if it came along. All his cancellation really means is that I need to find a new partner as quickly as possible if I'm going to exercise over my lunch break. Or maybe I should go for a jog instead, and then the question is whether my running shoes are still in the trunk of my car.

These sex differences in friendship networks and the relative weighting of autonomy and connection are then magnified by the enormous differences in circulating testosterone between men and women. Testosterone has wide-ranging effects on human behavior, one of which is to orient us toward competition. Higher testosterone levels lead to greater competitiveness and success in competition leads to a spike in testosterone, creating a self-perpetuating cycle. Because men have much higher testosterone levels than women, they are more competitive than women and expect other men to be the same. As a consequence, male relationships are more tolerant of competitive behavior than female relationships, and indeed are often built around competition as the glue that keeps them together. For many men, the perfect friend is someone who shares their love of a particular sport and their ability level, as that combination allows them to compete as fiercely as possible without knowing in advance whether today will be a win or a loss.

Larger circles of looser friendships give men certain advantages, which are primarily experienced in terms of tolerance of competition and forgiveness of betrayal. Because men expect less from their friendships, they worry about them less and are less likely to pore over the details of a conversation to determine what so-and-so meant when he said such-and-such, with the result that their friendships bring them less stress. But, at the same time—and this is a biggie—men's friendships are also less supportive and rewarding than women's friendships.

In the latest survey of friendship in America, only 21 percent of men reported they receive emotional support from their friends and only 30 percent share personal feelings or problems with their friends. Furthermore, more than two-thirds of men don't have friends to turn to when they are in need. This lack of supportive friendships is problematic for a host of reasons, as it leaves men unmoored and disconnected from society, an outcome that is bad for individuals and the collective. In contrast, 41 percent of women receive emotional support from their friends and 48 percent share personal feelings or problems with their friends. Although the story is clearly worse for men than it is for women, it's worth keeping in mind that the picture isn't great for women either, half of whom do not share their personal feelings or problems with any of their friends. This long-term decline in friendships is a topic we'll return to in Part III of the book. For now, I'd just like to note that family and spouses presumably make up for some of the loss that men (and to a lesser extent, women) experience in missing friends. But neither is a perfect surrogate, as family and spouses have their own expectations and demands that often require us to avoid airing our true concerns or, in some cases, the depths of our despair. Which brings us to our next topic.

Sex and Mental Health

Women are twice as likely as men to be depressed and this effect can be seen as early as age twelve (the sex ratio in depression is more like three to one if we focus on early adolescence). This sex difference in depression is not just a function of who seeks help, as it also emerges in representative surveys in countries around the globe. Although we can't rule out the possibility that men are refusing to admit they're depressed, sex differences in depression are actually larger in countries that have less strict gender roles, suggesting that men and women are telling us how they really feel. The picture gets more complicated, however, when we consider the starkest consequence of depression: suicide. On the one hand, women are nearly three times more likely than men to *attempt suicide*, which matches the depression data. On the other hand, men are nearly four times more likely than women to *die by suicide*, and these findings from the US are essentially replicated in every country where we have good data.

These incommensurate sex differences in depression versus suicide raise important questions. Higher levels of male suicide in the context of fewer male suicide attempts are often attributed to the fact that men use deadlier means when they attempt suicide. And indeed they do, as men are more likely to shoot themselves than women are. Men are also more likely to shoot other people than women are, so the temptation is to dismiss higher male suicide rates as evidence of men's greater aggression, whether aimed outward or inward.

But the data don't line up well with that interpretation. Men are also more likely to hang themselves or poison themselves with carbon monoxide (from their car exhaust), the latter of which does not seem like a particularly aggressive means of death. Additionally, when we examine sex differences in suicide rates among people who use the exact same methods, men are still more likely to die by suicide. These data suggest that suicidal men are simply more determined to end their life, a sign of greater distress.

These complex data raise an obvious question: Why are women more likely to be depressed but men more likely to die by suicide? A common answer to this question is that society expects men to be self-reliant, so it's more difficult for men to seek help. This explanation certainly seems true, but why does society expect greater self-reliance in men than women? I believe the answer lies in the greater autonomy of men and the greater connection of women. Autonomy is a form of self-reliance in that it reflects independence from others. Connection, in contrast, is the antithesis of self-reliance as it reflects interdependence with others. Men want and expect independence, with the result that it's difficult for them to seek help when they need it. Women want and expect connection, with the result that it's easy for them to seek help. As we saw in the discussion of friendship, men's friendship networks also make it more difficult for them to get help and support even if they're inclined to seek it.

Self-reliance is a virtue among men in hunter-gatherer societies just as it is in industrialized countries, suggesting that our ancestors probably showed similar sex differences in autonomy and connection that had the potential to lead to similar sex differences in distress. Sex differences in autonomy were less problematic for our ancestors than they are for us today, however, because their connections were so strong. Modern men suffer from an absence of friendship that was simply never the case for our ancestors (a point we return to in later chapters). If men evolved to seek autonomy, then the solution to male suicide is unlikely to be found in efforts to push men to be less self-reliant and more willing to seek help. Rather, the solution to male suicide is likely to be found in strategies that help men recover lost connections from our past without impinging on their self-reliance.

By way of example, here in Australia we have an organization called Men's Shed that is designed to reduce isolation and loneliness among men.* The motto of Men's Shed is that *Men don't talk face-to-face, they talk shoulder to shoulder.* This approach to male friendship highlights that men aren't seeking direct emotional intimacy; men are about bonding while doing. Men come to Men's Sheds to work on projects for themselves or their community with a variety of tools that are kept on hand. There is a growing research literature around Men's Sheds that suggests they help reduce the sort of problems that are associated with despair, such as excessive alcohol consumption and depression. The advantage of the approach exemplified by Men's Sheds is that it makes it easy for men to get together while teaching or learning a new skill (or otherwise increasing their self-reliance), and in so doing, form much needed bonds with each other.

Sex and the Workplace

One of the biggest sex differences in preferences concerns interest in people versus objects. Women are more interested in careers that

---

* It has now spread to a number of other countries and they're always keen to start more if you're interested in getting involved.

involve working closely with others while men are more interested in careers that involve manipulating objects (for some men that means swinging a hammer, for others programming a computer). Similarly, women are more likely to find satisfaction in their work when they can help others, whereas men are more likely to emphasize personal success and self-improvement at work. These female interests are a clear expression of connection goals and these male interests are a clear expression of autonomy goals.

So far so good. Men and women enjoy different jobs and there are lots of jobs that need doing, raising the possibility that sex differences in interests will lead to a sex-based division of labor that makes everyone happy. The problem emerges, however, when sex differences in interests become intertwined with sexism in the not-so-distant past, making it difficult to disentangle when occupational sex differences reflect preferences versus discrimination. This problem is then ratcheted up considerably when sex differences in preferences, a history of sexism, and differences in remuneration all coincide.

The most notable example of this confluence of forces lies in the debates surrounding female participation in math and science (often referred to as STEM: Science, Technology, Engineering, and Math). Do women face a hostile climate that limits entry into these lucrative fields, as they did when I was a student? Or might sex differences in autonomy and connection predict the differential participation in STEM that we still see today? New data keep adding nuance to this question, so with the caveat that what I say next might be dated by the time you read it, here's what I see as the most likely answers to these questions.

Over the last decade, I've become convinced that the sex differences we still see in STEM are largely a product of sex differences in interests rather than inequity or differences in ability. Why do I make this claim? Consider the following three findings:

1. The ratios of men to women in the different STEM disciplines have been carefully documented for over fifty years and the relative ranking of those disciplines—from most to least male-dominated—has remained largely unchanged since the 1970s, even as women's participation has tripled and quadrupled. These relatively unchanging rankings, in the context of dramatically increasing female participation, raise the possibility that sex differences in career choice reflect the inherent attractiveness of different fields to men and women. Consistent with this possibility, the fields with more male representation focus on objects while the fields with more female representation focus on people. Although these sex differences could reflect differential discrimination by field, to accept that possibility requires us to accept that differential discrimination rates have remained largely unchanged across the different fields for fifty years. Furthermore, because the percentage of female engineers is lower today than the percentage of female psychologists was in the seventies, the differential discrimination explanation would also require us to believe that engineers are more sexist now than psychologists were in the 1970s. And that strikes me as wildly implausible. I remember the seventies well—cool clothes and fun music but ghastly attitudes.

2. Among people who have the excellent quantitative skills necessary for a career in STEM, those with excellent verbal skills are less likely to pursue STEM fields than those with ordinary verbal skills. These data suggest that people who are good at things other than science typically don't want to be scientists even when they're also good at science. Importantly, among men and women who have good quantitative skills, women are much more likely to also have good verbal skills. Men who are really good at math are often not much good at anything else, but women who are really good at math tend to be smart

all around. The implications of these data are provocative—
it's not that women are getting pushed out of STEM, it's that
men are getting pushed into it (because men who are good at
science have few opportunities elsewhere).

3. Women in countries with relatively low gender equity, such as
    in the Middle East and South Asia, are *more* likely to pursue
    careers in STEM than women in countries with relatively high
    gender equity, such as in Scandinavia. If discrimination were
    preventing women from entering STEM fields, we would ex-
    pect the exact opposite. After all, there must be more barriers
    for women to overcome if they want to be scientists in Algeria
    than in Finland, but women make up over 40 percent of the
    STEM graduates in Algeria and only 20 percent in Finland.
    These data are confusing, in that we might expect no effect
    for gender equity, but it's surprising that the effect is opposite
    what we would predict. The likely story here is that gender
    equity doesn't really matter, but money does. Countries with
    worse gender equity tend to be poor and careers in STEM are
    one of the clearest routes to financial success in any country.
    Among women who live in countries that are relatively poor,
    those who have good quantitative skills enter the sciences to
    ensure they can make a living. Among women who live in
    countries that are relatively rich, however, those who have good
    quantitative skills pursue less lucrative careers because they
    can do so without risking a life of poverty.

These data raise the possibility that sex differences in STEM
no longer have much (or possibly anything) to do with discrimination and barriers, but everything to do with sex differences in
connection and autonomy. If so, the millions of dollars that are
spent each year trying to encourage girls to enter a career in the

sciences might be better spent elsewhere. For example, nursing is even more gender biased than computer science (albeit in the opposite direction)—perhaps coding camps for girls could be replaced by caring camps for boys.

Finally, there is another side to the gender/STEM debate that gets little attention but where we see sex differences that are even larger. The fact that boys *tend* to outperform girls in STEM has led to a great deal of interest and concern, but the fact that girls *always* outperform boys in verbal tasks has garnered almost no interest at all. Parents are always keen for their children to start talking, but despite their universal encouragement, boys lag behind girls. For example, in one study in which toddlers were recorded at home interacting with their parents, half of the girls could string two words together by the age of fourteen months (e.g., "eat cookie") but none of the boys could. We all eventually learn to string two words together, but these sex differences in verbal ability never go away.

To get a sense of this effect later in life, we can compare sex differences in reading, math, and science scores among high school students in eighty different countries from Albania to Vietnam, all of whom are taking the same test in their own language. In every single country, female students outperform male students in verbal tasks. The math/science data are a bit bouncier, with females sometimes outperforming males and sometimes not, but there's no exception to the rule with verbal skills. Furthermore, in most countries these sex differences in verbal skills are larger than the sex differences in math/science.

Why do girls outperform boys verbally in every country on Earth? There are many answers to this question, focusing on brain structure, hormones, and life experience. But they all intersect at the same place, which is that females are more connected than males and connection, in turn, relies on verbal communication. Just as an emphasis on autonomy makes men more likely to pursue a career in STEM, an emphasis on connection makes women more likely to pursue communication-related careers. Good verbal skills may not be as lucrative as good math/science skills (at least not in today's world), but they play a critical role in almost all aspects of life. People who can clearly articulate their positions and preferences are in a much better position to have their needs met than people who cannot.

Beyond these complex issues encapsulated in the gender/STEM debate, sex differences in autonomy and connection have the potential to emerge in various other domains at work. To choose but a single example, women's greater emphasis on connection means that when women do compete with one another, they typically pay a steeper relationship cost than equally competitive men. As discussed in the section on friendship, women are relatively intolerant of competition and conflict in their relationship partners, which means they have less latitude to engage in autonomous or competitive behavior with other women. Because the workplace often demands autonomy, women can find themselves in the dicey situation of choosing between upsetting their friends or upsetting their boss. Men are largely freed from these office dramas, as they expect their friends to be autonomous and competitive, so they don't really care when inevitable workplace conflicts arise.

Unlike what happens in Vegas, what happens in the workplace doesn't stay in the workplace; sex differences in the workplace invariably leak into people's private lives. An emphasis on autonomy leads men to put the brunt of their efforts into the workplace, as they strive for success and recognition. For many men, success at work is one of the key ways they see themselves taking care of their family and meeting their connection goals. Women typically do not feel this way. They are less concerned about the workplace (even when it matters to them), particularly once they have children, as they are more focused on connection. One result of this discrepancy is that women do way more housework and childcare than men do, even when both are employed full time.

Unfortunately, the obvious differences in who is doing what around the house can overshadow the fact that men put in a lot more hours at work than women do, even when both are employed full time. For example, there's a wonderful longitudinal study of highly gifted men and women that tracked them from adolescence into their fifties. Men in this sample spend about ten more hours per week in the office and women spend about ten more hours per week on housework. The sample isn't representative of the general population, as the men and women in it are mostly higher earners. But because women in the sample have so much earning potential, we know that if they chose to spend a few extra hours at work, their salary could easily pay for a few extra hours of childcare and housekeeping. Furthermore, even though the sample isn't representative, the findings are: the work patterns we see with this highly gifted sample are similar to what we see with other Americans.

It's always possible that the highly gifted men had more rewarding jobs than the highly gifted women, for any of a host of reasons, leading them to spend more time in the office. However, when these men and women were asked at age fifty how many hours they'd be willing to work for their ideal job, fewer than 10 percent of the men but 25 percent of the women were *unwilling* to work full time even for their ideal job. This sex difference in willingness to work full time is reflected in other preferences held by these men and women. For example, the men in this sample report a greater interest in earning a high income and being successful at work while the women report a greater interest in spending time with family and having close relationships. These data show us that the sex differences that emerge in what people actually do are reflected in what they want to do.

Unfortunately, these sex differences can be a source of friction, even though couples typically agree that both parties are working equally hard and even though both parties tend to regard the distribution of labor as fair. The problem lies in the fact that both parties also agree that men *want to put in extra hours at the office* but women *don't want to do the dishes or the diapers.* The issue here is that doing housework is supportive of women's connection goals but not a direct expression of them. Dishes and diapers are necessary chores for communal living, but neither activity brings you closer to others. Indeed, women (and men) usually enjoy the time they spend taking care of their family, just not the time they spend doing chores.

In contrast, extra hours at work are a direct manifestation of men's pursuit of autonomy goals, as they lead to greater productivity and competence and hence a more autonomous self. This inequity is then compounded by the fact that men take on fewer parental duties, in large part due to their reduced connection needs. These sex differences in outside work, housework, and parenting loom large in many families, causing both sides to feel unappreciated by a partner who has different views on the balance between autonomy and connection.

Sex and Leisure

The differences between men's and women's choices in leisure activities follow the same patterns we've seen in friendship, mental health, and work. Men enjoy leisure activities that focus on autonomy while women enjoy connection. To start with reading material, men's favorite books are likely to be about struggles to define yourself and your place in society, like *The Catcher in the Rye* and *The Outsiders*, or struggles against the world, like *Catch-22* and *1984*. Women read more books than men (fourteen versus nine per year), are more likely than men to read fiction, and their choices differ as well. Women's favorite books are more likely to be about relationships, with *Jane Eyre*, *Pride and Prejudice*, and *Wuthering Heights* high on women's list but not on men's list at all.

These differences are also reflected in women's and men's choices of magazines, which have varied over time (*Good Housekeeping* and *Ladies' Home Journal* are no longer the powerhouse publications they were in the fifties and sixties, nor are *Adventure* and *Man's Life*), but which show the same divide. The magazines that interest men and women today are more egalitarian in their depiction of gender roles, but the same themes emerge. Women are more likely to read *Better Homes & Gardens*, *Family Circle*, and *People*, while men prefer *Sports Illustrated*, *Popular Mechanics*, and *Field & Stream*. Here again we see a predominance of stories about people and connection versus objects and skills—the same sex differences we see in occupational preferences.

Movies and television show the same pattern, with TV programs like *James & Mike Mondays* (videogames) and *Fifth Gear* (automotive) appealing to a predominantly male audience, while *When Calls the Heart* (a drama about a single teacher) and *The Lizzie Bennet Diaries* (a modern instantiation of *Pride and Prejudice*) appeal to women. When men do prefer shows about relationships, they are of the dysfunctional variety, such as *Beavis and Butt-Head* and *Rick and Morty*. At movie theaters, women make up the brunt of the audiences for rom-coms and musicals while men fill the seats in action, sci-fi, and horror movies. Finally, musical tastes reflect similar sex differences, although bands and genres often defy easy categorization. Iron Maiden and Rage Against the Machine appeal to men more than women, while Taylor Swift and Miranda Lambert appeal to women more than men. Sex differences in musical preferences go beyond sex differences in autonomy and connection, as can be seen in the fact that songs about relationships are popular with both men and women. But the emphasis on relationships is more evident in the music of Taylor Swift than Iron Maiden and what they sing about relationships differs as well (compare "Labyrinth" to "22 Acacia Avenue").

The Conflict Caveat

So far, the data have all lined up rather nicely: women show greater connection than men at work, home, and in their leisure activities, while the reverse is true for autonomy. There is one notable exception to this widespread pattern, however, in which men are more connected than women (and hence more likely to sacrifice their autonomy). Indeed, the connection that men feel for each other in this one domain is every bit as strong as the tightest connections we ever see among women. When are men tightly connected to each other? During *intergroup conflict*.

As I've already mentioned, the greatest threat to humanity has always been other groups of humans, as our capacity to work together is unparalleled in the animal kingdom. Furthermore, the mortal aspects of intergroup threat are directed primarily at men. When groups of hunter-gatherers come into conflict with one another, they typically show no mercy to males from the other group. Females from other groups don't fare well either, as they are vulnerable to capture and subsequent slavery in their new group, but over time they can eventually obtain full rights of group membership in most societies.* These rules of conflict are so pervasive that we often see sex-based signatures of ancestral conquests in the genetic remains of ancient men and women. Time and again, archaeologists discover different origins for ancestral male and female lineages, with most or all the men suddenly being replaced by another group while the female lines become absorbed rather than disappearing altogether.

---

* Women who are kidnapped often develop strong loyalties to the groups that captured them, despite the horrors they and their loved ones experienced during the conflict that led to their capture. For example, early American settlers were often shocked to discover women of European origin living in Native American tribes who were not interested in being "rescued."

Given the regularity and ferocity of ancestral intergroup conflict, male psychology evolved in numerous ways to protect men from genocide. The most notable adaptation is that men bond very tightly with each other whenever they're in conflict with other groups. Intragroup bonding might seem like an odd solution to the threat of annihilation, but it has always been the best approach to an otherwise intractable problem. By bonding with other men in their group, human males became the most effective fighting force this planet has ever seen. Male-male bonding not only leads group members to work together and defend each other more effectively, it also makes them more willing to go into battle as they know their comrades have their back. For these reasons, evolution favored men who formed strong attachments to one another in times of conflict.

This aspect of male psychology is very old—probably older than our species—and it still manifests itself in numerous ways. First, when you talk to soldiers, particularly members of units who have seen a lot of combat together, they speak of the extraordinary closeness they feel for each other when they're in battle. Many of the commandos I have chatted with have told me that when they are at war, they feel closer to each other than they do to their own families. This tight bonding is one of the goals of training, but it emerges most strongly among units who have survived live fire together and it slowly dissipates as the fighting abates.

Second, it's not only soldiers who experience this form of male bonding. The feeling may not be as strong, but modern sports teams create a similar mentality, particularly contact sports in which team members defend one another from bodily harm. American football is a prime example, as everyone's job on the team is to protect each other from members of the other team, who are intent on mayhem.

Team members bond to each other even when there's no threat of bodily harm, but there's nothing like knowing that your teammates are saving you from being crushed into the turf to connect men to each other. This effect is most powerful for the athletes who experience it directly, but they're not the only ones who bond over intergroup conflict. Fans also experience an incredibly strong connection to their group when they watch competitive sports, and contact sports heighten that feeling. Despite all the physical damage wrought by the game of football,[*] and hence all the reasons not to play, it remains one of the most popular pastimes in America. Nearly every high school has a football team and the most popular television programs continue to be football games (in 2022, eighty-two of the top one hundred shows were NFL games). The NFL is more popular among men than it is among women, with 52 percent of American men saying they are avid fans and 24 percent of women saying the same thing. You need only watch men in the bleachers during the game to realize that male bonding suffuses the sport, from the players to the spectators.

When you reflect on these sex differences in preferences, it's clear they have numerous sources. Big effects like these are almost always multicausal, in the sense that it's not just one key factor like testosterone or estrogen that leads people to feel the way they do. Cultural practices play a substantial role in these sex differences, as the media, our friends and family, and numerous other sources of information communicate appropriate norms to men and women and enforce those norms via various pressures, subtle and otherwise. These cultural effects stand out most clearly when we look at changes over time, as women are much less interested in traditional gender roles today than they were in the 1950s.

But numerous lines of evidence suggest that culture isn't the

---

[*] Consider Damar Hamlin's near death in the game between Buffalo and Cincinnati or the chronic traumatic encephalopathy (CTE) brought about by repeated head injury that has led to suicides and murder among ex-football players. Current estimates suggest that over 90 percent of ex-professional players suffer from CTE.

whole story. First, the effects we've been discussing in this chapter emerge in almost every culture for which we have good data, despite the enormous differences in gender egalitarianism across cultures. Indeed, one of the most striking effects we find in cross-cultural research is that sex differences are often magnified in cultures with *more* rather than *less* gender egalitarianism. We discussed this idea with regard to the gender/STEM debate, but the same holds true in other domains, such as sex differences in altruism and the tendency to reward kindness in others (F>M) versus risk-taking and the tendency to punish unkindness in others (M>F). Women everywhere are more likely than men to reward kindness, but this sex difference is larger in Scandinavia than it is in the Middle East. In contrast, men in every country are more likely than women to punish unfriendliness, but again this difference is larger in Scandinavia than it is in the Middle East.

Second, when we're able to collect relevant data, we see similar sex differences in other great apes, who obviously aren't influenced by the media or other sources of culture in the ways they maintain relationships, attempt to dominate others, and even in their toy preferences when they are young. Finally, while many of these sex differences have shrunk over time, they have not disappeared altogether. That finding could simply be evidence that not enough time has passed, which means we need to keep a close eye on the data for the next few decades at least. But for now, it's worth noting that the size of many of these sex differences has remained constant for at least the last twenty to thirty years, directly after a few decades of rather steep decline in their magnitude.

We interpret these data as evidence that sex differences are likely to have a biological origin, but it's worth remembering that cultural practices can also be highly stable and deeply ingrained. As we'll see in the next chapter, cultural differences play a major role in the weighting that people give to their autonomy and connection needs.

5

Connection to the East, Autonomy to the West

Does the squeaky wheel get the grease or does the nail that sticks out get hammered down? When we reflect back on our lives, most of us would say the answer is both,* but the odds of getting what you want versus getting flattened also depend on where you live. American culture is highly individualistic, meaning that individual preferences (autonomy) often take precedence over other people's expectations (connection). Americans encourage their children from a very young age to develop and express their preferences and they expect the world to accommodate them. In contrast, people in most other parts of the world live in collectivist cultures, meaning that individual preferences often take a back seat to the expectations of others. People in collectivist cultures learn they must accommodate themselves to the world, rather than the other way around. This differential weighting of autonomy and connection in individualist and collectivist societies can be seen in countless ways, from the substantial to the trivial. Consider the following questions:

---

*Small children having a temper tantrum in the checkout line strike me as the classic example. Sometimes they get a snack, sometimes they get a smack.

1. If you were offered a promotion that required you to move far from friends and family, how much of a raise would it take for you to move?

    If you declined the promotion to stay in your hometown, would your friends and family be surprised? Would they think more or less of you?

2. If you were out to dinner with colleagues and the two people who ordered ahead of you both chose the dish you were considering, would you stick with your order or switch to a different one?

    If you're tempted to switch your order because your colleagues just ordered the same thing, why? Were you planning to eat off their plate? How does their choice impact your own?

3. If you were competing with a friend in one of your favorite sports or games and won the competition after a string of losses, would you trash talk with your friend or say that you were just lucky?

    If you did trash talk, would your friend be amused or annoyed?

Your answers to these questions reflect your upbringing, personality, sex, etc., but they also reflect where you grew up. Americans move at the drop of a hat; collectivists require a greater incentive before leaving friends and family behind. Americans feel awkward if they order the same things as their colleagues; collectivists often feel comfortable when they do the same thing. And Americans (particularly American men) love to trash talk in competition, but collectivists are much more circumspect when they have beaten a friend.

In this chapter we'll focus on the underlying meaning of these differences and how members of different cultures emphasize connection versus autonomy in their orientation toward each other and their world. But first let's take a peek at the sources of these cultural differences. Collectivism and individualism emerge in some places and not others for a reason, and it will help us understand these cultural habits if we have a sense of their prehistory.

What Are the Origins of Individualism and Collectivism?

When we consider the life of hunter-gatherers, it's clear that all human societies were once highly collectivist. Because our ancestors relied so heavily on one another to survive and thrive, their connections to each other were paramount. Those connections, in turn, demanded that social responsibilities were weighted above individual preferences almost every time the two came into conflict. Our survival depended on cultural rules that created and enforced such responsibilities, with universal sharing of meat being the prime example. Nonetheless, the constant demands of our connections wear us down by impinging on our autonomy. It's hard to run your own show when you're intertwined in a dense network of mutual responsibilities.

This dynamic, in which connection is weighted over autonomy by virtue of necessity, leads to a clear prediction that has been borne out across the world: as cultures become wealthier, they slowly shift from collectivism to individualism. But wealth isn't the only thing that matters. Numerous other factors come into play when determining whether a society will be on the collectivist or individualist end of the spectrum. Before considering these factors, let's take a look at the world to see where different countries fall on this continuum. Thanks to the pioneering efforts of Geert Hofstede, one of the most influential researchers in cross-cultural psychology, we have data on collectivism and individualism from most parts of the globe.

The first thing you notice when you look at Hofstede's data is that Australia and New Zealand, North America (sans Mexico), and Western Europe are individualist outliers in an otherwise collectivist world. Their outlier status is all the more apparent when you consider that Australia, New Zealand, and North America are populated primarily by immigrants from Western Europe. These data suggest that individualism is a minority enterprise with roots in Western Europe, while the rest of the world is collectivist.

What caused individualism to take root in Western Europe? Or, asking the question another way, what caused collectivism to stick around so long everywhere else? It's tough to know the answers to these questions with certainty, as so many historical factors are intertwined, but we can see a few factors at play. If we begin with hunter-gatherers, we see a particular form of collectivism that reflects their unique lifestyle. On the one hand, the risk of relying on hunting for food meant that hunters were required to share the fruits of their labor, which is a strong form of collectivism. On the other hand, their nomadic existence meant that if they disliked others in their camp or disagreed with group decisions, they could either splinter into subcamps or head off to join another group (both of which were common). The freedom to associate only with people you like and accept only the rules with which you agree are strong forms of individualism, although both were limited by people's ability to find enough like-minded others with whom they could form a camp. Going it alone was simply not an option, so our ancestors' autonomy was always constrained by their need to be part of a collective.

The shift to farming eliminated mandated sharing, as people could now grow and store their own food and thus were not subjected to the vagaries of animal movements and other factors that could stymie even the most gifted hunters. Because success was now more tightly linked to hard work and less tightly linked to luck, the rules requiring people to share outside their families largely disappeared when they started farming. That's a potential win for individualism, but it would take many generations before people's obligations to their clan became any less onerous than the requirement of sharing the spoils after a successful hunt.

The shift to farming also reduced the opportunities to associate with whomever you'd like and avoid rules that don't suit you. Once people were living on a piece of land they had cleared and planted, they weren't leaving no matter how obnoxious their neighbors might be or how annoying the local council's rules were. For these reasons, the early days of agriculture were less individualistic than the hunter-gatherer life our ancestors had left behind, but farming paved the way for individualism.

The type of farming that people chose—or that the local ecology supported—also impacted whether individualism emerged in different farming communities. To return to our question of why individualism originated in Western Europe, one answer can be found in the fact that farming in Western Europe was centered on grains such as wheat and barley. These cereals can be farmed without much help from others beyond the immediate family. In contrast, rice farming requires community-level involvement at key points in the process, lest the crops fail. Rice farming is also so water intensive that it depends on extensive irrigation, which requires farmers on neighboring plots to cooperate with one another to share the available water and maintain the interconnected irrigation systems. Consistent with these differences in the cooperation demanded by rice versus wheat farming, the southern areas of China, which traditionally farmed rice, are more collectivist than the northern areas of China, which traditionally farmed wheat.

Western Europe was also heavily influenced by the dictates of the Catholic Church, which created institutions and cultural rules that broke up the traditional power structures of family and clan. For example, marriage among first and second cousins was very common in Europe (and elsewhere) prior to the advent of the Catholic Church, but subsequently forbidden by Catholic Law. As a result, people in Western Europe began to look outside the family for their life partners, which weakened the hold clans had on daily life, reducing the responsibilities people had to one another as a function of clan membership.

In addition to these differences in farming and religion, Western Europe is also the seat of the Industrial Revolution, a set of changes in technology and practice that led people to become self-sufficient even while they remained embedded in collectives. Factories depended on the combined output of a large number of workers on the factory floor, but employees took home their paycheck as a function of their own individual performance. In this manner, industrial workers became individually self-reliant, even if their employers still depended on large-scale cooperation to generate the desired outcomes.

Finally, cultures also vary in the extent to which they are "loose" or "tight," or the degree to which members have the latitude to behave as they want. All cultures have rules, but loose cultures have lower expectations than tight cultures do that people will actually follow those rules. For example, it's against the rules pretty much everywhere to pull open the subway door if you arrive as it's closing—at that point you're meant to wait for the next train. Despite my full knowledge of that rule, I've jammed my hand or foot into a NYC subway door several times when the doors started to close in my face. On those occasions, the doors bounced back, I hopped on, and no one cared or even bothered to look up.

When I tried that once in Japan, not only was I the recipient of aghast expressions from every single person in the car I hopped into, but I also froze the entire rail network. It turned out that the software in the train didn't allow the doors to be pulled back open without freezing the entire system, *because no one ever does that*. When every single one of the 125 million Japanese people arrive as the train door closes in their face, they wait for the next one. Alas, not I, leading me to spend the longest fifteen minutes of my life under the withering glares of my fellow commuters as the railway staff ran through the enormous station individually restarting every door in the entire train before it could move. By the time we were underway again, the line to board the train stretched out of the station and I was pretending to be invisible. That experience engraved upon my brain the fact that systems in tight cultures are simply not designed to withstand casual rule breakers.

As is the case with collectivism more broadly, necessity is the cause of cultural tightness. Cultures that have a long history of natural disasters, wars, famine, and other threats, as well as high population density, tend to be tighter than cultures that are more sparsely populated and relatively protected from such threats. If life is risky or people are packed in like sardines, cultures become very tight to manage the challenges of living there. If life is safe and people are sparse, cultures can afford to be much looser. Western Europe certainly has its crowded cities and dangerous places, but on average it has been less crowded and more protected from natural disasters than Asia or Africa. Although cultural tightness is distinct in principle from individualism/collectivism, tight cultures tend to be more collectivist than loose cultures for the obvious reason that the demands of tight cultures require greater conformity and norm compliance from their members.

So far we've been discussing cultures as if they're monolithic entities, but it shouldn't surprise you that people vary within cultures as well as between them. We spent a fair bit of time on that basic point in Chapter 4 when we talked about sex differences. The consequence of this inevitable variation is that we don't all fit our culture of birth equally well; some of us have the good fortune of being a perfect fit, whereas others feel like square pegs being hammered into round holes. This state of affairs has implications across a variety of domains. For example, fitting in with your culture is associated with higher self-esteem, greater social support, and better mental and physical health. Indeed, our match with our culture is one of the factors that drives migration, as people are more likely to pull up stakes and move along when they don't fit in. This mismatch effect appears to be stronger among people who are more focused on connection and weaker among people who are more focused on autonomy. This finding makes perfect sense, given that people who focus on connection will struggle more if they don't fit in, whereas people who focus on independence are less likely to notice (or care) if they stand out.

It's also important to keep in mind that in the same way that all individuals have both autonomy and connection needs, all cultures have individualist and collectivist components. When you travel to other cultures, however, you tend to notice the differences more than the similarities, and they're often confusing if you don't know what they mean. I've had the good fortune of taking a few round-the-world voyages on the Semester at Sea program, which allows you to see a big slice of the world in a leisurely fashion and also lets you see the world through the eyes of your fellow travelers. Misunderstandings in both directions are all too common on such trips, but one example that stands out in my mind was when our ship's doctor gave his customary lecture on health concerns as we approached Vietnam. We had just left China, and he mentioned as an aside that he was surprised that the Chinese were so worried about catching colds, as he had noticed a fair few people wearing face masks as he walked around Shanghai (this was in the pre-COVID world of the 1990s, when no one considered wearing masks as a matter of course).

What obviously hadn't occurred to him is that in a collectivist culture, it's bad manners to go out in public with a cold without masking up to reduce transmission to others. As an individualist, he assumed their masking behavior was an effort to protect themselves from other people's respiratory illnesses. But as a collectivist, the same behavior can easily be understood as an effort to look out for others. Fast-forward to the COVID pandemic, and these differences in the tendency to mask up out of responsibility for others were readily apparent. Collectivist countries were faster to adopt mask mandates than individualist countries, and their citizens were more compliant when the mask mandates went into effect. There were even regional differences in masking rates in China that matched the different farming practices, with rice-farming areas showing more rapid adoption of masking than wheat-farming areas. The degree to which masks ultimately reduced the spread of COVID remains a matter of debate, but the point is that people in collectivist countries were more focused on following norms and rules and less concerned about their individual rights.

The doctor's confusion about masking was a clear example of an individualist interpretation of collectivist action. It was also trivial in comparison to some of the misunderstandings people experience as they travel. Consider, for example, the simple acts of saying yes and no. *Yes* is easy to say everywhere, but in collectivist cultures, *no* is far more complicated than it might seem. People have no trouble saying no to factual questions, like "Is today Tuesday?," but saying no to an interpersonal request is not so easy when people feel strong responsibilities for each other. In many collectivist cultures people go out of their way to avoid directly declining requests. Tepid or vague forms of agreement are polite ways of saying no that confuse individualists, who either think they got no answer at all or that they got a yes. Nine times out of ten when students and friends on the Semester at Sea program told me of plans that went awry when they were socializing with locals, it was clear that someone had tried to say no to them but had done so in a manner that was too subtle for their individualist ears to hear.

Cultural Practices in Impression Management

One of the hardest things we humans do is manage the impressions we make on others. Nothing is more important than our reputation because nothing has a bigger impact on how others treat us. Given the emphasis on fitting in and conformity in collectivist societies, it should come as no surprise that collectivists are particularly unimpressed when people try to enhance their reputation by tooting their own horn. No one likes a braggart, but collectivists like them even less. "Don't brag" seems like a simple enough rule, but it does leave us in a bit of a quandary when things go our way and our friends and family don't know. How do you reap the social benefits of success without paying a reputational cost for telling people?

Let's start with social media, which is one of the major venues for selling yourself in the modern world. We expect our friends' posts on Facebook to be highly curated, with photos of beach vacations and gourmet dinners rather than subway commutes and fried bologna sandwiches.* So we have no problem when people put up pictures of fun things and ignore the drudgery; indeed, that's what we want and expect them to do. This form of bragging is only noticeable when the vacations get a bit too fancy or the expensive dinners a bit too frequent.

But what should you do when you achieve something notable? How do you let the world know you got into Harvard? Or, if you're like me and Harvard wasn't an option, how do you let people know you won second place in the fourth-grade spelling bee?† One solution that works in the face-to-face world is to wait for people to ask. Or better yet, wait until they bug you so relentlessly that

---

*Keeping in mind that there are Facebook countercultures in which people post their vending machine snacks and dirty subway rides. For every group norm there's a counternorm out there somewhere.

† Which I did: Rogers Park Elementary, Mrs. Wacker's class, 1972.

you have no choice but to admit your accomplishments ("Alright already! Yes, I heard from Harvard, and somehow I tricked them into accepting me . . ."). Of course, they might not ask. And if they do, they might not pursue the topic if you're vague in your initial responses. Either way, no one is asking you anything on Facebook, Instagram, or LinkedIn, so you need to find a way to tell the world without looking like a tool.

The most common approach to this problem is along the lines of "I was humbled and honored to receive the . . ." This opening line is perfectly acceptable; people know their major successes rely on help from others, with a bit of luck thrown in, and hence it's true that they're both humbled and honored when they receive individual recognition for their achievements. The problem with this approach is that it's also a transparent effort to tell the world about your achievements when you're worried they might not find out otherwise. You're much better off if you can somehow nudge your friends into spreading the news for you, and better off still if despite your apparent efforts to bury the signal, it somehow leaks out. Our ancestors didn't have this problem, as everyone saw them dragging the giraffe into camp at the end of a successful hunt. But our achievements in today's world tend to be much more difficult to discern, which creates a challenge when we want to share the good news.

This problem is difficult for all of us, but it's a real bear for people from collectivist societies. To brag in such cultures is to expose yourself to ridicule. From what we can tell, this rule has been in place since we were hunter-gatherers. Recall that highly skilled hunters need to downplay their achievements even more than ordinary hunters, lest their campmates become jealous or worry that the best hunters will try to dominate others. Downplaying achievements is still the norm in collectivist societies: the more successful you are, the more you downplay it. Because everyone knows that such modesty is a form of politeness, it enhances your reputation to claim to be less than you are. This rule also holds among individualists, but it works better among collectivists, who know that modest claims are a common cover for success. Of course, some people who claim to have failed really have, but if people know that modest claims are required from particularly high achievers, they quickly learn what these claims mean as a function of who makes them.

The Modern Shift Toward Individualism Has Been a One-Way Street

Over the past fifty years, cultures around the world have slowly drifted toward individualism. Many cultures show little or no change in this regard, but the ones where we can see clear evidence of change are universally shifting away from collectivism. We've already discussed the role of economics in this shift, as money enables independence, but there are other factors at play as well. By way of example, let's consider China's one-child policy, which was intended to prevent China from becoming overpopulated.

There are numerous interesting and unintended consequences of the one-child policy, but for our purposes we can focus on changes in parenting practices that have followed from it. In the forty-five years that have elapsed since the policy was implemented, China has become a country in which not only do most parents have only one child, but most grandparents have only one grandchild. It's easy to see how that one kid can become all-important to the family, as the child is the focus of all their hopes and aspirations as well as the recipient of all their attention and affection. If you share but a single grandchild with three other grandparents, how often do you think you're going to remind her of responsibilities and duties to others? How often will you reprimand her when she is selfish or thoughtless, as children often are? Not often. The one-child policy has created a world of "little emperors," in which the previous cultural emphasis on roles and responsibilities has shifted to an emphasis on what will make your child or grandchild happy and successful.

If we extrapolate from China to the rest of the world, it becomes clear that shrinking family sizes all over the world are associated with increasing individualism. As people have smaller families, they cater to their children's wants more (and presumably remind them of their responsibilities less). When you have fewer children or grandchildren, the temptation to spoil them gets stronger and more achievable. With the average family now having fewer than two children across the industrialized world, we have another reason why autonomy is ascendant.

These demographic changes in the US are well documented, so they serve as a useful example. When I was a child in the 1960s, the average family had nearly four children and the two parents put in a combined total of less than two hours per day in direct childcare (in case you're curious, 90 minutes from Mom, 21 minutes from Dad). Fifty years later, American families have approximately two children and parents are putting in slightly more than three hours per day in childcare (117 minutes from Mom, 65 minutes from Dad). Thirty minutes of parenting per kid per day in the 1960s has tripled to 90 minutes per kid per day in the 2010s. I'm not claiming that's 60 additional minutes of child-focused positive attention (what you might call spoiling), but when I look back at my childhood, I have very few memories of my parents playing children's games with us except when we were on vacation. When I reflect on my own parenting, my primary recollection is of sore knees and back as I got down on the floor to play with my toddlers.

Finally, along with increasing wealth and shrinking family size, there's a third reason why the world is becoming more individualistic: collectivism is hard on happiness. As we consider how this process works, keep in mind that cultural practices don't evolve to make us happy; cultural practices emerge, take hold, and spread when they make people more effective in their struggles with the environment or each other. But people like to be happy. As cultural practices that lead to unhappiness become less necessary for survival, later generations are often disinclined to adopt them and they slowly disappear.

Why might collectivism cause unhappiness? If connection to others is our most fundamental need, how could a society that emphasizes connection make its members unhappy? There are three answers to this question. One of them we've already discussed extensively, which is the cost to autonomy that people pay whenever they emphasize connection, so we can tick that off our list. The other two answers emerge from the fact that connection entails responsibilities, and the more our connection is driven by fundamental survival needs, the more important those responsibilities are. By way of example, imagine a person who is trying to become a great basketball player, either because she thinks it's fun or because she dreams of joining a school team.

If this person (we'll call her Kristen) is playing out of her own interest—for reasons of autonomy—she is beholden to no one but herself. If she plays well or poorly, her ability or lack thereof benefits and costs no one else. For that reason, no one will really care if she doesn't practice regularly and no one will be upset if she misses a shot when she's playing at the gym. It's also the case that no one will be happy to see her practicing at all hours, nor will people be excited to see her hit shots from every angle and distance. In contrast, if Kristen is playing for the school team—for reasons of connection—she has a responsibility to others. Her failure to practice is a direct affront to her teammates (we'll call them the Hawkeyes), who are depending on her not to let them down. If she misses a shot or block, her failure will cost the whole team, but if she makes a great shot or block, her success will benefit everyone. All the Hawkeyes are invested in the performance of connected

Kristen, particularly when it's a big game, but no one particularly cares about the performance of autonomous Kristen.

This is the precise situation people face when they are members of collectivist versus individualist cultures. Recall that collectivism is a psychological response to an environment that demands cooperation if people are to survive. We saw this with hunter-gatherers and we saw it with rice farmers. Hunter-gatherers and rice farmers are members of collectives, or teams, that really matter. If I'm to survive as a hunter-gatherer, I count on you to succeed on some of your hunts so you can share with me when I come home emptyhanded. And if I'm to survive as a rice farmer, I count on you to maintain my irrigation ditches once they cross into your land and to work closely with me at critical points in the harvesting process. For these reasons, I care a great deal about the work you're doing because it has a huge impact on me. In contrast, if I'm a wheat farmer, what you do with irrigation on your land is largely irrelevant to me and I don't need your help to bring in the crops, as there are no periods of such high-intensity work that I can't handle the job myself.

And this brings us back to the two remaining reasons why collectivism can make people unhappy. First, if I'm highly dependent on you for my own success, I'm going to engage in two interrelated behaviors: I'm going to be nosy and I'm going to engage in social comparison. If some of your hunts are going to feed me, I want to know how you're preparing your arrows the night before and whether you have a good plan for where you're going to hunt today. I'm also going to pay close attention to what you bring home at the end of the day and what you fail to bring home. Similarly, if the way you maintain the irrigation channels on your land is critical to the survival of my own rice crops, I want to see that you're doing your job properly.

As a result, I'm not only going to attend to what you're doing, but I'm also going to compare your performance to everyone else's. Otherwise, it can be ambiguous whether your success or failure is a product of luck versus skill. But if you're the only hunter who hasn't caught anything in the last few weeks, or if you're the only farmer whose irrigation system isn't delivering water after the recent rains, your failure is on you. So, your success or failure is not the only thing I care about; I also care how it stacks up to everyone else's, including my own. This process of social comparison helps us understand how others are doing and it tells us how we're doing, too. Because we know that everyone else is engaged in this same comparison process, it doesn't take a genius to realize that if other people are reliably outproducing me, I'm soon going to find myself shunned by my group.

In short, collectivism increases the extent to which people engage in social comparisons with others as well as the importance they attribute to the outcome of these comparisons. Because we're social beings, we'll always care if we're outperformed by others (particularly if we're competing for mates), but being outperformed has more dire consequences for collectivists than for individualists. Furthermore, collectivists are more likely to notice if they're being outperformed, because they're paying so much attention to everyone else to be sure that others are doing their part.

And this takes us to the final reason why individualists are happier than collectivists: criticism. There's no point in paying attention to the job you're doing and comparing it to others if I'm not going to do anything about it when I discover that your performance is inadequate. The reason I'm paying so much attention is to ensure that you do your job right so I survive. As a consequence, when I encounter poor performance on your part, it's incumbent on me to intervene and explain your inadequacies to you. You may or may not want to know why your irrigation ditch failed, but if I have opinions on those issues, you can count on the fact that I'm going to offer them. Furthermore, you can also count on the fact that our cultural rules will dictate that you not only hear me out, but that you do your best to improve your performance.

In contrast, if I'm a wheat farmer and you mosey on over to tell me why you think my farming practices are subpar, I can listen or ignore you at my discretion. I'm not beholden to you, and no one else in my community particularly cares if my crop yield is high or low this year. One consequence of this difference is that rice farmers and other collectivists learn to be more self-critical than wheat farmers and other individualists, as collectivists are more motivated than individualists to prevent failure and subsequent criticism from others. Better to find your own failures before others do, particularly when you know that others will.

The outcome of these two processes is that collectivists engage in more social comparisons than individualists do, those comparisons matter more to collectivists, and collectivists are subjected to more criticism from self and others when their efforts don't stack up. Social comparison and criticism are not necessarily a bad thing— particularly if you need advice—but they can be hard on your psyche. If you have any experience with cognitive behavioral therapy (also known as CBT), you'll know that it's one of the most effective forms of talk therapy for the treatment of anxiety and depression. There are many aspects to CBT, but one key thing it teaches us to do is change the way we talk to ourselves, in part by reducing self-criticism. Through CBT, people learn not to blame themselves for everything that goes wrong and not to catastrophize or magnify the harms that might emerge from their behavior or situation.

It doesn't take a huge leap to see that collectivists are at greater risk of depression and anxiety, given that their mental lives often directly oppose what CBT would recommend. And that, in turn, brings us back to where we started, which is that the road from collectivism to individualism has been a one-way street, with individuals and societies shifting steadily toward autonomy and individualism when their circumstances allow it.

The world was once a highly collectivist place. Collectivism has enormous pluses but it's not easy to be a collectivist. As we've become richer and safer, collectivism has become less necessary and most cultures have responded by dropping collectivist practices. This shift in our emphasis from connection to autonomy made perfect sense when we were hyper-connected hunter-gatherers or early farmers. As we'll see in Part III, however, there are numerous other forces at play that have moved us toward autonomy and away from connection, with the result that the pendulum has now swung too far. Before we turn our attention to this problem, however, we need to consider the roles played by religion and politics in how each of us draws the line between autonomy and connection.

Religion Redefines Autonomy and Connection

Over the last two thousand years, Abrahamic religions (Judaism, Christianity, and Islam) with an all-powerful and moralizing God have captured the hearts and minds of more than half of humanity, defining what most of us mean when we speak of religion. But religion meant something very different to our ancestors, who usually believed in many gods rather than just one, and whose gods were neither moralizing nor omnipotent. For example, our ancestors usually venerated the dead, believing that their direct ancestors became gods after death, capable of influencing their current fate on Earth in matters large and small. If you've ever seen football players pray before a game or thank Jesus after a touchdown, you get a sense of the sort of personal requests our ancestors made and the gratitude they felt toward their ancestors when things went their way.

Our ancestors also believed that animals, plants, and often the earth, sky, and cosmos, had different spiritual essences that one could pray to for luck, favor, or relief. Like us, our ancestors knew their prayers would not always be answered, but they had a different understanding for why their pleas might be ignored. Sometimes their gods were paying attention and sometimes they weren't.

Sometimes their gods had the power to grant wishes and sometimes they didn't. Perhaps most importantly, sometimes their gods punished wicked behavior, sometimes they didn't, but their gods were not overly concerned with how people treated their neighbors, much less their enemies. For example, when Joseph Watts and his colleagues looked for evidence of moral punishment in the earliest known religions of the South Pacific and Indian Ocean regions, only six of ninety-six religions had omnipotent gods who reliably monitored and enforced moral codes. Our ancestors' gods were not terribly concerned with how we treated people who were not members of our extended family.

Religion Broadens Connection

The Old Testament, also known as the Jewish Bible or Torah, predates the Christian Bible and Quran and is the origin of the Abrahamic religions. In some ways it represents a contrast from prior religious tradition and in some ways it does not. The Old Testament breaks sharply from the past in its emphasis on morality, with numerous laws intended to enhance cooperation and minimize conflict. If we take the Ten Commandments as representative of the basic approach laid out in the Old Testament, we see that the first four concern proper ways to worship (no other gods before me, no graven images, do not take the Lord's name in vain, honor the Sabbath) but the remaining six focus on moral concerns that ensure a smoothly running community (with the mandate to honor your parents and prohibitions against killing, adultery, stealing, lying, and coveting). The religions of our ancestors had few such moral guidelines.

There are many reasons why religion began to mandate moral behavior as we moved from nomadic hunting to agriculture and cities, but one important factor is that people who lived in the same community were no longer interdependent or necessarily even known to each other. Strangers who don't need each other (city dwellers) are less likely to look out for one another than people who encounter each other regularly and depend on each other (hunter-gatherers). Cooperation tends to break down when people are strangers, when their interactions are one-off, when self-interest is not tied to the interests of others, and when people are concerned that others might take advantage of them. All these situations emerged with our shift away from a hunter-gatherer lifestyle.

By providing a set of moral guidelines that are enforced by an all-knowing and all-powerful God, the Abrahamic religions incentivized cooperation. Hunter-gatherer religions didn't concern themselves with morality because cooperation was already enforced by their cultural rules and interdependent living conditions. Once we abandoned the lifestyle of hunter-gatherers, our cultural rules no longer mandated universal cooperation, with the result that new rules were needed to keep people connected. According to this perspective, religion provided the social glue that facilitated connection when our daily existence stopped doing so.

If we set aside this emphasis on morality, we see that the Old Testament shows a fair bit of continuity with prior religions in the way it is steeped in the sort of mayhem that was common in small-scale societies. There are numerous examples of Old Testament violence to choose from, but nothing beats the story of Moses' battle with the Midianite people, in which his troops killed all the men. When they returned with the women and children as captives, Moses was angered by their compassion, directing his army to "Kill every male among the little ones, and kill every woman who has known man by lying with him. But all the girls, who have not known man by lying with him, keep alive for yourselves" (Numbers 31:17–18). Moses' demands could have been taken directly from the hunter-gatherer conflict handbook, although hunter-gatherers were less concerned with whether their female captives had known a man by lying with him.

To our modern sensibilities, Moses' genocidal instructions are deeply disturbing, but we need to keep in mind that morality evolved to guide the treatment of one's own group. As we've discussed earlier, when we were hunter-gatherers, interactions with other groups might be an opportunity for trade or finding new romantic partners, but other groups were often a mortal threat. In the latter case, ruthlessness was the most reliable survival strategy; moral concerns have no place in the treatment of outsiders who would show no mercy if they gained the upper hand.

Despite recommending wholesale slaughter when in conflict, the Old Testament offers a philosophy of intergroup kindness in times of peace. Consider, for example, Leviticus 19. Most of this book is composed of laws that facilitate cooperation within one's own group—an agriculturist's guide to the sort of mutual aid provided by hunter-gatherers—but 19:33–34 lays down a law that few hunter-gatherers would recognize:

> If a stranger lives as a foreigner with you in your land,
> you shall not do him wrong. The stranger who lives
> as a foreigner with you shall be to you as the native-
> born among you, and you shall love him as yourself;
> for you lived as foreigners in the land of Egypt.*

In this manner, the Old Testament took an important step toward broadening our connections to outsiders. Even bigger changes to the Abrahamic religions were afoot, however, with the New Testament (the Christian Bible, which is not endorsed by Jews but accepted by Muslims) preaching a form of kindness that was previously unknown. As we move forward in time to the New Testament, we see

---

* This remarkable advice is sandwiched between *Show respect for the elderly* and *Do not use dishonest standards when measuring length, weight, or quantity* (scales were invented about five thousand years ago and were rife with error in biblical times, so it was pretty easy to cheat your customers via inaccurate measurement).

Jesus doubling down on a law from Leviticus in the Old Testament (19:18 Thou shalt not avenge, nor bear any grudge against the children of thy people, but thou shalt love thy neighbour as thyself) when he relates the parable of the good Samaritan (presented here in slightly abbreviated form).

THE GOOD SAMARITAN (LUKE 10:25–37)

An expert in the law asked Jesus, "What shall I do to inherit eternal life?"

"What is written in the Law?" Jesus replied.

LAWYER: "Love thy neighbor as thyself."
JESUS: "You have answered right."
LAWYER: "And who is my neighbor?"

[Note: This is the key question here—who are we being asked to love?—Leviticus 19:18 is pretty clear that we are meant to love our own people.]

JESUS: "A man went down from Jerusalem to Jericho, and fell among thieves, which stripped him, wounded him and departed, leaving him half dead. And by chance there came a priest that way, and when he saw him, he passed on the other side. And likewise, a Levite. But a certain Samaritan saw him, and went to him and bound up his wounds, brought him to an inn, and took care of him. Which of these three thinkest thou was a neighbor unto him that fell among thieves?"
LAWYER: "He that showed mercy on him."
JESUS: "Go and do likewise."

Much of the meaning of this narrative has been lost in the ensuing centuries, but Samaritans were a despised minority and Jesus was emphasizing one of his key points that *everyone* is deserving of your compassion, not just your own people. This concept might seem so obvious today that it's hardly worth mentioning, as our moral circle now encompasses all of humanity (or at least those components of humanity with whom we're not currently in conflict). But as we've seen, it wasn't always this way. In the parable of the Good Samaritan, Jesus was asking people to reject the dominant cultural norms by connecting with people outside their ethnic group.

Perhaps just as remarkably, Jesus also took a giant step toward defusing conflict by attempting to short-circuit the endless cycle of revenge that was common prior to the establishment of effective policing. Recall that Jesus was born into a world in which the Old Testament's "an eye for an eye" was the law of the land. In case you're not familiar with the original text, it's worth noting that this rather harsh rule applied even in cases of unintended harm (here it is in its original form, Exodus 21:17–25).

Now suppose two men are fighting, and in the process they *accidentally* strike a pregnant woman so she gives birth prematurely. If no further injury results, the man who struck the woman must pay the amount of compensation the woman's husband demands and the judges approve. But if there is further injury, the punishment must match the injury: a life for a life, an eye for an eye, a tooth for a tooth, a hand for a hand, a foot for a foot, a burn for a burn, a wound for a wound, a bruise for a bruise. [emphasis my own]

In his Sermon on the Mount, Jesus explicitly rejects this law and asks his listeners to turn the other cheek. Mahatma Gandhi, Martin Luther King Jr., and their followers are some of the only humans I've ever heard of who had the self-control to follow Jesus' request, but the fact that it was probably intended as aspirational advice makes it no less remarkable.

Numerous scholars have argued that humans domesticated themselves, in part by executing members of their own group who were overly aggressive. Such self-domestication via selective execution would have created its own evolutionary force. Because highly aggressive people met an untimely death, the genes that predisposed people to high levels of aggression and violence would eventually be depleted from the gene pool. According to this possibility, by repeatedly enacting certain patterns of behavior that altered our evolutionary landscape, we changed our own genetic makeup. In contrast, Jesus' arguments for broadening our connections in the New Testament are a form of societal change sometimes referred to as cultural evolution, in which cultures change when new ideas take hold and replace old ways of doing things. The people are the same, it's the rules that differ.

Sometimes genes and culture interact in such changes, for example when the control of fire led to cooked food and eventual genetic changes that reduced the size of people's jaw muscles (as large muscles were no longer required to chew food that was softened by fire). But often cultural changes simply result in new reward structures that disincentivize behavior that was previously useful. My favorite example of such cultural evolution can be found in Scandinavia. Despite being genetically indistinguishable from their Viking ancestors, Scandinavians no longer make a living via raiding and pillaging. Indeed, Scandinavians are among the most peaceful people on Earth. Because robbing and killing one's way up and down the coastline is not a good way to make a living in modern Europe, Swedes have given up this practice. So have Norwegians and Danes.

In his Sermon on the Mount and the parable of the Good Samaritan, Jesus was asking for the sort of change we've seen in Scandinavia over the last twelve hundred years. If everyone could agree to love their neighbor (a prior mandate from the Old Testament), and if everyone could further agree that everyone is their neighbor, human compassion would be universal. Just as importantly, if people could stop seeking an eye for an eye (another mandate from the Old Testament) and instead turn the other cheek, the world could avoid the endless cycle of revenge that the political philosopher Thomas Hobbes felt was one of the major causes of violence.

This expansion of our moral circle from our family and immediate group to encompass all of humanity is an enormous ask—you need only consider how far we still are from achieving it—but the key is that Jesus made this request in all seriousness. This is a form of connection that didn't exist previously, that still doesn't exist fully, but that nonetheless describes how a substantial number of us feel about all other humans.* Perhaps this emphasis on universal connection is why women tend to be more religious than men when they are members of moralizing religions like Christianity, despite being less religious than men when they worship non-moralizing deities and spirits.

Religion Deepens Autonomy

Not only did the New Testament change the way we connect, but it redefined autonomy as well. First and foremost, the Old Testament emphasizes action over thought. Of the hundreds of commands instructing people how to go about their daily lives in the Old Testament, fewer than 5 percent concern thoughts. Rules about thoughts exist—most notably against coveting in the Ten Commandments—but they are comparatively few and far between. The Old Testament is much more concerned with maintaining proper behavior, and hence a smoothly running society, than it is

---

* If you doubt this claim, consider that 25 percent of American households contributed to relief efforts after the deadly 2004 Asian tsunami. Eighteen years later, 25 percent of Americans again contributed to victims of a foreign disaster, this time in support of Ukrainians after the Russian invasion. One quarter of a population contributing their hard-earned cash to people in a country they couldn't find on a map is pretty impressive.

with maintaining an unimpeachable mental life. I suspect this emphasis on a smoothly running community is why Exodus demands "an eye for an eye" even in cases of accidental harms. A system of rules in which even accidents have serious consequences will lead people to be very careful in how they treat each other. Nevertheless, there are reasons to focus on thought as well as action—if you start to dwell on forbidden fruit, it can be very hard to resist.

Consider the prohibition against sex outside of marriage, where the data suggest that inappropriate thought and behavior are exceedingly common. The Old Testament provides two separate prohibitions against extramarital sex in the Ten Commandments alone, with Commandment Seven "You shall not commit adultery" and Commandment Ten "You shall not covet your neighbor's wife." This law against coveting your neighbor's wife, only a few sentences after a law banning adultery, suggests that prevention of sexual jealousy was of paramount importance for keeping the peace.

In his Sermon on the Mount, Jesus wanted to be sure we understood that even in the absence of inappropriate behavior, inappropriate thought is still problematic: "You have heard that it was said, 'You shall not commit adultery; but I tell you that everyone who gazes at a woman to lust after her has committed adultery with her already in his heart" (Matthew 5:27–28). Moments earlier in his sermon, Jesus makes a similar point regarding murder, telling his listeners they will be judged by God not just for committing murder, but also for being angry or speaking ill of others.

How do rules prohibiting certain thoughts change the meaning of autonomy? On the one hand, a God who can see our innermost thoughts and consign us to an eternity of torture would seem to negate the very idea of autonomy, as such an entity can demand complete obedience and all but a fool would comply. On the other hand, such a God could easily reveal himself to everyone in the sort of miraculous fashion that would ensure all are convinced of his presence and power, yet he does not do so. Christians understand that his decision not to reveal himself publicly in all his glory is intended to give them the *choice* of whether to believe and follow him. The God of the New Testament is asking us to choose whether we want to take the Path of Righteousness and save our eternal soul. From that perspective, the God of the New Testament has given us more autonomy than we ever had before. It's hard to imagine leaving something more important than eternal bliss versus eternal torture to the individual to decide.

The New Testament also changes the meaning of autonomy itself in its emphasis on autonomous control of thought as well as action. Consider the following experiment by Christina Starmans and Paul Bloom, who presented children and adults with stories in which a child was asked to do something that either did or did not conflict with her other goals. For example, in one story the child was asked to clean up her toys. Half the time the story indicated that she would rather play with her friends and half the time it did not, but either way the child always complied and cleaned up her toys. When the participants were asked which of the two children "deserved an award for doing something good," three-quarters of the children preferred the unconflicted protagonist but three-quarters of the adults preferred the conflicted protagonist. These data suggest that children regard following the rules as more virtuous when your heart is in the right place, but adults understand that some rules are particularly hard to follow and give people more credit when they follow the difficult ones. That is, adults recognize that sometimes our *wants* conflict with our *oughts* and it's particularly impressive when our *oughts* dominate our *wants*.

That all seems clear enough, but it turns out it's not that simple. After all, the New Testament tells us that thoughts themselves can be immoral. Consider the following experiment by Adam Cohen and Paul Rozin: in their research they examined differences between Jews (who only follow the Old Testament to the exclusion of the New) and Protestants (who endorse both Bibles but emphasize the New Testament) by presenting them with the following scenario:

> Mr. K. has never liked his parents very much. Mr. K. finds them to be too involved in his life and they have very different personalities and goals from him.

The scenario then went on to say one of the following:

> Nevertheless, Mr. K. has always made sure to behave as if he really cares about his parents. Mr. K. is sure to call and visit his parents regularly. Every year, he sends his mother flowers on her birthday and his father a box of cigars on his birthday. When his parents grow old, they know that they will not have to worry about anything because Mr. K. will take care of them.

Or:

> Because Mr. K. does not like his parents, he oftentimes forgets to call them for a few weeks at a time or to send them gifts on their birthdays. To tell the truth, Mr. K.'s parents feel as if he neglects them. When his parents grow old, they do not know that Mr. K. will take care of them.

Cohen and Rozin found that Jewish participants rated the helpful son much more positively than the neglectful son. These data suggest that the Jewish participants give Mr. K. credit for looking after his parents despite not liking them. In contrast, the Protestant participants were relatively insensitive to Mr. K.'s behavior and rated him negatively for not honoring his parents in his heart, regardless of whether he looked out for them.

Here we see an interesting difference from the Starmans and

Bloom findings. In Starmans and Bloom, children respond positively to people who *think* the right thing, but adults give people credit who overcome an internal conflict to *do* the right thing. In Cohen and Rozin, Protestants give no credit to people who overcome an internal conflict to do the right thing *when the internal conflict is caused by morally unacceptable thoughts*. This might seem odd if you're not a Protestant, but we all do this to some degree when we encounter people who have sufficiently immoral thoughts. For example, if you learn that someone desires to have sex with animals, or worse yet, with children, you are likely to feel revulsion for him even if you know he never acts on his desires. You might be willing to give him some credit for his restraint, but you will assuredly be appalled by his preferences.

What do these experiments tell us when we consider them together? The data suggest that religion has changed our understanding of autonomy to the point that Christians (particularly Protestants) place a great deal of weight on people's thoughts and not just their actions. In this manner, the New Testament has led to a broadening of autonomy to include decisions regarding the contents of our own minds. Earlier I defined autonomy as "self-governance, choosing a path based on your own needs, preferences, or skills, and making independent decisions." This definition takes thoughts as the starting point; people display autonomy when their behaviors are driven by their own preferences rather than those of others. Nevertheless, in principle we can apply the same rules to thought as to behavior, by asking whether people allow or suppress their thoughts based on their own needs and preferences or the needs and preferences of others.

This focus on internal states is related to a second key aspect of Protestantism that has changed the meaning of autonomy, which is the notion that salvation is individually earned. Ancestral religions typically emphasized following the proper rules to please or placate the spirits; follow the rituals and you have your best chance of getting what you want. The rituals and their purposes changed with the advent of Abrahamic religions, but Jews and Catholics continue to emphasize ritual in their religious practice, as well as completion of these rituals with other members of their community. Faith matters, but it doesn't take precedence over ritual and community.

In contrast, in the five hundred years since Martin Luther declared that people can relate directly to God without involvement of the Church, salvation for Protestants has been based largely on their individual relationship with God. Faith is much more important than practice or community to Protestants. Connection with co-religionists is a critical component of religion for Catholics and Jews, but autonomy and self-direction are more central to the religious experiences of Protestants.

Due to their focus on autonomy, Protestants are more likely than Catholics to see people's behavior as driven by internal qualities such as personality rather than external qualities like situational factors. This increased emphasis on autonomy means that Protestants give people more credit for their successes, but they're also more inclined to blame them for their failures. We'll take a deeper dive into politics in the next chapter, but one difference between Democrats and Republicans (and Left and Right more generally) lies in their emphasis on individual responsibility. Democrats see a variety of barriers to success (based on one's race, sex, sexual orientation, social class) that can stymie an individual's best efforts, while Republicans view success as available to anyone who has the talent and puts in the effort.

Because Protestants are more focused on autonomy than Catholics and Jews, we would expect that Protestants are also more likely to be Republicans than Catholics or Jews. The data are consistent with that prediction: Protestants are the most likely to be Republicans, Jews are the least likely, and Catholics fall in between. There are many reasons for people of different religions to identify with different political parties, so this finding alone is not very convincing. But when we focus more closely on people's attitudes regarding individual success, the story remains the same. For example, Protestants are the most likely to believe that government aid to the poor does more harm than good, Jews are the least likely, and Catholics again fall in between. A person who perceives systemic barriers to success would be unlikely to endorse such an opinion regarding government aid, but a person who believes hard work leads directly to success would want to incentivize hard work by minimizing other options. Protestants are the most likely to explain success and failure as a product of individual effort.[*]

Not by Prayer Alone

So far we've discussed how religion changed the meaning of both connection and autonomy, by broadening the former (to include all of humanity) and deepening the latter (for Protestants, at least, by making their personal choices the determinant of their own eternal salvation and by making them responsible for their thoughts as well as their actions). It should come as no surprise that religious experience also reflects connection and autonomy differently for different branches of the Abrahamic religions. First and foremost, religious practice is a highly social act for many if not most adherents. But, as noted, Protestantism emphasizes autonomy, whereas Catholicism and Judaism emphasize connection. Thus, we would expect the social benefits of religion to be more evident for Catholics and Jews than for Protestants. To test this possibility, I used the General Social Survey to examine the

---

[*] People often wonder how religious individuals could vote for candidates whose personal behaviors are counter to religious teachings, but data such as these show that religion can focus attention on factors that only indirectly relate to the religious teachings themselves.

happiness levels of a representative sample of Americans of these three religions.* To examine the effects of different types of religious participation, I compared the happiness of people who never prayed to those who prayed more than once a day, the happiness of people who never read the Bible to those who read it daily, and the happiness of people who never attended services to those who attended services more than once per week. Everyone in the sample I examined self-identified as either Protestants, Catholics, or Jews.

Each bar in Figure 6.1 represents the effect of a religious activity on happiness, as each bar shows the difference in happiness between people who never do the activity versus people who do it frequently. As you can see, members of all three religions report

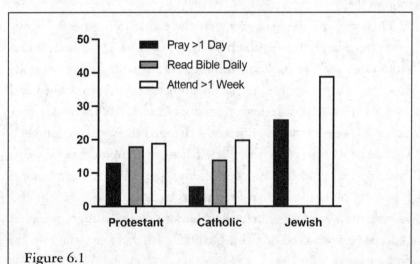

Figure 6.1
Increase in percentage of Protestants, Catholics, and Jews who are made very happy by their religious activities (compared to those who never do these activities).

---

* The General Social Survey is an extraordinary, publicly available dataset that I'll be returning to throughout this book. It can be analyzed online at https://gssdataexplorer .norc.org.

greater happiness when they participate more extensively in their religion. We don't understand this effect of participation very well, but we'll return to it in Chapter 9. In the meantime, let's take a closer look at the figure to see how different aspects of participation affect members of these three religions.

To start with Protestants, we see that the increase in happiness associated with regular religious attendance is barely larger than the increase associated with regular Bible reading or prayer.* For Protestants it doesn't make much difference whether they practice their religion alone or with others. In contrast, among Catholics the increase in happiness associated with religious attendance is substantially larger than the increase associated with reading the Bible or prayer. We see a similar story among Jews when we have enough data.

These polling data suggest that the connection aspects of religion play a larger role in the happiness of Jews and Catholics than Protestants and are probably a more central part of their religious experience. Consistent with this possibility, when Adam Cohen and Peter Hill asked Protestants, Catholics, and Jews "Have you ever had an experience that significantly changed the way you approach life or the world?," they found that life-changing experiences differed in their emphasis on connection across these three religions. Jews were more likely than Protestants to report a social event and Protestants were more likely than Jews to report a direct personal experience with God. Again, Catholics fell between the two on both answers.

These data suggest that the most important experiences of Jews and Catholics are likely to focus on connection with other people, while the most important experiences of Protestants are likely to

---

* The survey didn't ask whether they pray or read the Bible alone, so we can assume that sometimes these activities are social, but neither is likely to be as reliably social as attending religious services.

focus on their individual relationship with God. In so doing, they remind us that while changing societal concerns about connection are reflected in religious doctrine, religion also guides these shifts in meaning. Sometimes religion pushes most of us in the same direction, for example, when the three Abrahamic religions exhort us to open our hearts to people from distant lands. But religions push us in different directions when it comes to the relative weighting we place on autonomy and connection and even in what these needs mean to the individual. Given how fundamental connection and autonomy are, it should come as no surprise that members of different religions often fail to understand each other. As we'll see in the next chapter, mutual misunderstanding is also common across political lines, for many of the same reasons.

7

Connection to the Left, Autonomy to the Right

Liberals and Conservatives view the world from different vantage points. People on the political Left (Liberals) tend to be most concerned with fairness, harm, and intergroup relations, all of which emphasize connection. People on the Right (Conservatives) are typically more concerned with individual rights, governmental overreach, and excessive taxation, all of which emphasize autonomy.* Connection is also important to those on the Right, but particularly in the form of connection to one's group, which manifests in concerns for loyalty, patriotism, and duty. Similarly, autonomy remains important to those on the Left, but more so in the ways that the autonomy of disadvantaged groups may be hindered by structural barriers to achievement.

Although people on the Left and Right share all these values, they differ in which values they prioritize, leading to substantial divergence in their policy preferences. By way of analogy, you might

---

* As in Chapter 4, from here forward I'm going to drop the "tend to" and "typically," in that I'll be discussing average differences between the Left and Right, not how any individual Liberal or Conservative feels.

think of people on either side of the political spectrum as compet-
itors in the game of bridge who are all playing together but who
disagree about which suit is trump (the trump suit dominates
or outranks all other suits when played). For people on the Left,
fairness and avoidance of harm are trump values, and all other
values are secondary in importance. For people on the Right, individual rights are trump values, and even connection is secondary
(although, as we'll see, not by much). Conflicts between the Left
and Right arise primarily when people are forced to sacrifice some
of these values in favor of others.

To get a sense of how differential weighting of values plays itself
out in political opinions, let's start with US polling data. The trade-
off between taxation and social spending is a great entry point;
not only does it capture a fiscal reality that social spending needs to
be paid from tax revenues, but it also encapsulates the fundamental
tension between autonomy and connection. If I take money from
you in the form of taxes, I'm impinging on your independence by
preventing you from deciding how to spend your own money. Because money buys options, I'm also removing some of your flexibility when I tax you. Taxation reduces autonomy. If I then spend your
tax dollars on social programs (welfare, food stamps, job training
programs), I'm addressing your concerns for others by improving
the lives of people in your community. Social spending fulfills connection goals. Taxes also pay for bridges, bureaucrats, and bombs,
but spending on social programs (including Medicare, Welfare, and
Medicaid) is approximately one-third of the budget, meaning that
a third of our tax dollars are spent helping others.

How do people feel about the trade-off between autonomy and
connection that underlies taxation for social spending? To examine
this issue, we can go back to the General Social Survey and look at
polls in which people are asked if the US government should reduce taxes or increase social spending. As you might guess, people's
answers depend on whether they are on the Right or the Left.

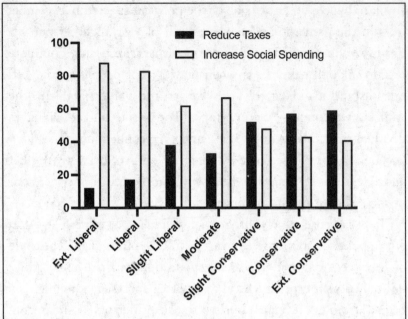

Figure 7.1
Percentage of people who believe it's more important to reduce taxes or increase social spending as a function of political orientation.

As you can see in Figure 7.1, the trade-off between autonomy and connection depends on political orientation. Extremely liberal people prefer to increase social spending while extremely conservative people prefer to reduce taxes. Before we dive deeper into this effect, two things are worth noting. First, within neither party do we see complete consensus. Even among extremely liberal people, for whom connection is paramount, 10 percent would rather reduce taxes than increase social spending. And even among extremely conservative people, for whom autonomy is their trump card, 40 percent would prefer to increase social spending over reducing their taxes.* Second, despite these huge differences in pref-

---

* In all of these agree/disagree graphs, I've removed people who neither agree nor disagree, so the totals don't always add to 100 percent.

erences between people on the far Left and far Right, connection beats autonomy overall. People on the far Right are four times more likely to choose connection (by spending money on people they don't even know) than people on the far Left are to choose autonomy (by reducing their own taxes). And that, in turn, explains why fully a third of your tax dollars redistribute your money to people in need—most Americans prefer it that way.

One potential concern with the differences between Left and Right in these data is that they are sullied by self-interest. People on the Left tend to be poorer than people on the Right, so perhaps their greater favorability toward social spending reflects an expectation that social programs are more likely to target themselves or their family and friends. Although that may be true, we see the same basic effect when we remove this possibility from consideration. For example, in Figure 7.2 pollsters asked whether rich countries should use their tax dollars to help poor countries, effectively

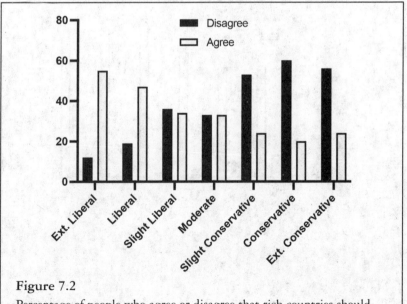

Figure 7.2
Percentage of people who agree or disagree that rich countries should use taxes to help poor countries as a function of political orientation.

ruling out the possibility that people on the Left might support social spending in hopes that their own family will benefit.

The effects of political party aren't as large here as they were for the social spending question, but the differences between the two parties remain stark. People on the far Left are more likely to endorse spending their tax dollars to support poor countries, while people on the far Right are more likely to disapprove of this use of their tax dollars. But again we see that connection beats autonomy. People on the far Right are twice as likely to agree to spend their tax dollars on poor countries as people on the far Left are to disagree.

So far, the data support the idea that people on the Left care about connection and people on the Right care about autonomy. But what about connection to your group? Recall that I argued that when people on the Right prioritize connection, it's in the form of patriotism and loyalty. People on the Left care about patri-

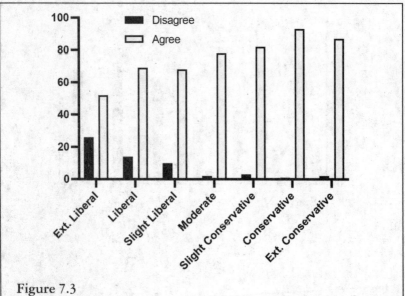

Figure 7.3
Percentage of people who agree or disagree that patriotism makes America strong as a function of political orientation.

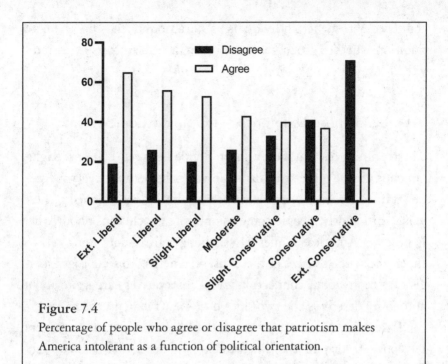

Figure 7.4
Percentage of people who agree or disagree that patriotism makes
America intolerant as a function of political orientation.

otism and loyalty, too, but that aspect of connection is less important to them than individual connection. We see evidence for these
claims in Americans' attitudes about patriotism. When pollsters
ask people if they think patriotism makes America strong, the answer is a consistent yes. But on the far Left that yes is endorsed by
a slim majority, whereas by the time we get to the far Right not
only does almost everyone agree, almost no one actively disagrees
(see Figure 7.3).

Why do some people on the Left disagree that patriotism is
good? We see the answer to this question when pollsters ask if
patriotism makes America intolerant. Two-thirds of people on the
far Left think patriotism is a source of intolerance while two-thirds
on the far Right disagree (see Figure 7.4). For people on the Right,
loyalty to one's group is paramount, making them unwilling to
consider that it might have a dark side. For people on the Left,

connection to other individuals is paramount, causing them to question whether group connection might lead to costs at an individual level.

Where Do These Differences in Opinion Come From?

These polling data suggest a sharp divide between Left and Right, which is all the more striking when you consider that everyone values connection and everyone values autonomy. By weighting these values differently, people come to opposite conclusions about what is best for America. Nonetheless, these polling data don't tell us much about the sources of these disagreements. To state that Liberals care more about connection and Conservatives care more about autonomy describes the two sides but doesn't explain them.

If we accept that autonomy and connection are in permanent tension with one another, then to explain the divide between Liberals and Conservatives we need only find something that pushes people in one direction or the other. Anything that would make Liberals care more about connection will also make them care less about autonomy, and anything that would make Conservatives care more about autonomy will make them care less about connection. As luck would have it, there are numerous reasons for both. Remember, large differences between people are usually caused by more than one factor, and these political differences are no exception to that rule.

Why are harm and fairness so important to the Left? Let's start with empathy, which is the predisposition to *share the feelings of another*. Liberals are higher in empathy than Conservatives* (if you want to see the data, take a look at the left-hand bars in Figure 7.5 on page 137). When people describe those on the Left as soft-

---

*With one rather ironic exception: Conservatives are more empathic toward Liberals than Liberals are toward Conservatives.

hearted or "bleeding-heart Liberals," they're referring to the fact that people on the Left are highly empathic. People who are high in empathy feel a real boost when others do well, which is a wonderful bonus for empathic people who get to benefit from the success of others. But there is also a cost to empathy in that empathic people feel bad when others suffer harm or unfair treatment. These bad feelings run deep. For example, children as young as six years old show larger metabolic, cardiac, and inflammatory responses to their parents' poverty and mental health problems if they're high rather than low in empathy. If you empathize with other people who are struggling, you struggle too. This inescapable price of empathy is why social justice is so important to the Left. Empathic people feel better when they address harms or unfair treatment of others because it alleviates their own vicarious suffering.

If vicarious suffering doesn't sound so bad to you, consider a series of famous experiments initiated by Dan Batson and his colleagues. In their research, participants were (falsely) told that the investigators were studying impression management under stressful circumstances. The participants were asked to draw straws to see if they would be subjected to electric shocks themselves or if they would be assigned to observe another person being mildly electrocuted. The drawing was always rigged to ensure the real participant was the observer. Participants then watched "Elaine" (an employee of the experimenter pretending to be a fellow participant) as the experimenter hooked her up to a shock machine. As she was being connected to the device, Elaine asked the experimenter how strong the shocks would be and was told "they would cause no permanent damage." The experimenter went on to say, "You know if you scuff your feet walking across a carpet and touch something metal? Well, they'll be about two to three times more uncomfortable than that."

Unfortunately, it turns out that poor Elaine had been thrown from a horse onto an electric fence as a child and was rather traumatized by the experience. As a consequence, she pretended to react rather badly to the first two shocks, getting noticeably upset. When the experimenter saw this, he asked Elaine if she'd like to continue. Elaine insisted she'd be fine and that she didn't want to ruin the experiment. At that point, the experimenter came up with the clever idea that perhaps the participant and Elaine should switch roles so Elaine wouldn't have to suffer and the experiment could continue. The question was what percentage of people would be willing to take the electric shocks themselves to end Elaine's suffering.

There was one more critical factor in these experiments that we need to consider before we see how people reacted to this offer. Half of the participants had been told they would need to watch Elaine get all ten of her shocks, but half had been told they could stop watching after two of them. So everyone could escape her suffering by taking it upon themselves, but half of them could also escape her suffering by simply skedaddling now that the first two were over. What did people do? It won't surprise you to learn there were lots of skedaddlers, but in a big win for human kindness, most people opted to switch places with her even when they could turn away and try to put her suffering out of their mind. The chances of turning away also went down when people felt more empathy for her.

What are we to make of these findings? First and foremost, vicarious suffering is real. More than half of the people in these experiments would rather take the electric shocks themselves than watch someone else squirm as she is (pretending to be) shocked. This effect was stronger among more empathic people, who were less likely to turn away and more likely to take her place. High empathizers couldn't bear the thought of her continuing agony, preferring to take on the unpleasantness themselves because they knew they could handle it (not having been thrown onto an electric fence themselves).

How do we know it was their inability to bear the thought of her suffering that led people to take her place? To test this possibility, Bob Cialdini and his colleagues replicated Batson's experiments, but gave half the participants a pill that would ostensibly "freeze" their moods, making it impossible to improve the way they feel for a short time. The pill was really just a placebo; the goal was to convince people that it would be hard to improve their mood so they shouldn't bother trying. Cialdini found that the pill made people less likely to switch places with Elaine and more inclined to simply walk away. When they thought they were stuck in their current mood, people were less inclined to help.

These findings suggest that part of the reason empathic people want to help is to reduce their own sadness and inner turmoil. These experiments also show us that when people feel empathy for someone else's pain, they feel real pain themselves—pain they're so motivated to alleviate that they'll suffer electric shocks just so they don't have to watch someone else writhe in agony. Consistent with this possibility, when we put someone in an fMRI machine and measure their brain functioning while they experience pain or while they witness someone else in pain, many of the same neural networks are activated in both cases. Vicarious pain hurts.

Where do these individual differences in empathy come from in the first place? We know genetics play a role, as genes explain about half of the reason why some people feel others' emotions and some don't. We also know that women are more empathic than men, a finding that is consistent with the sex differences in connection and autonomy we discussed in Chapter 4. Indeed, women's greater empathy is a large part of the reason why they're more likely to be Liberals and this sex difference in political preferences is not a small effect. For example, in the 2020 election, Trump got 53 percent of the male vote but only 42 percent of the female vote. That election might seem like cherry picking, given Trump's personal history with women, but a 10 percent gender gap is typical of what we've seen over the last thirty years. For example, in the 2012 election, Romney got 54 percent of the male vote but only 44 percent of the female vote. To the best of my knowledge, Romney's personal behavior with women has been impeccable. These data tell us that if men were not allowed to vote, there would have been no President Trump and if women were not allowed to vote, there would have been no President Obama. Empathy matters.

Empathy plays a central role in the differences between Left and Right, not just for the reasons discussed above, but also because empathy is wrapped up in other social attitudes. For example, the Left and Right also differ in their beliefs about whether people can be trusted and whether the world is a safe place. If I think people are essentially kind and trustworthy and the world is a safe place, I'm going to feel much more empathy toward others than if I think people are out to get me and the world is full of dangers. Furthermore, when I try to make sense of other people's behavior, I'm likely to give them the benefit of the doubt if I believe people are fundamentally good. Sure, they might have done something wrong, but in all probability they found themselves in dire circumstances and had little choice but to break the law. My fundamental belief in the goodness of others will make me forgiving of misbehavior, orienting me toward rehabilitation rather than retribution in my choice of punishment. Empathy drives connection, which will make me inclined to help people get back on the straight and narrow.

In contrast, if I think the world is a dangerous place and people are not to be trusted, I'm not only going to feel less empathic, I'm also going to look for a hidden agenda when making sense of people's behavior. Sure, they might have done something nice, but their actions were probably a prelude to some sort of evil plan. The only way to stymie such people is to remain on guard. When I do find them engaging in the sort of unpleasant behavior I'm expecting, the most sensible strategy is to respond with harsh retribution to prevent them (and others who might be watching) from taking advantage. It's mission critical in a dangerous world to show no mercy, as people will pounce at the slightest sign of weakness. Such attitudes are undoubtedly sensible under some circumstances, but they do not foster connection.

These differences in beliefs about human nature play a role in the perspective people take when they make sense of each other. When psychologists began studying how people understand each other, the metaphor they relied on might be called "perceiver as intuitive scientist." Psychologists thought of humans as dispassionate observers of each other as they attempt to discern underlying dispositions, which makes a fair bit of sense given how valuable the truth is. But there is more to the way we understand each other than our desire to get it right. One key factor is the cost of getting it wrong and whether that cost varies by the type of mistake you make.

If you can't be 100 percent certain that you're going to get it right, then you need to keep in mind that some errors have greater costs than others, and those are the ones to be avoided. By way of example, if you're hiking through the outback in my adopted country of Australia, and if you're aware that Australia is home to all ten of the world's most venomous snakes, you're likely to be alert to the possibility that a dangerous snake might cross your path. In such circumstances, you would prefer to err on the side of caution rather than nonchalance. Better to waste your energy jumping backward a hundred times when you see a stick that looks a bit like a snake[*] than to accidentally step on a single one of those top ten snakes.

The same psychology that causes us to be hyper-vigilant for venomous snakes works for our interactions with other people. Sometimes we're intuitive scientists, carefully trying to understand Sid's affinity for team sports when we hear him tell Richard that he really loves to bowl and is thinking of joining Richard's wife's league. But sometimes we're more concerned with detecting and preventing misbehavior. Is Sid trying to exploit Richard's trust and good nature, and if so, might he try to pull the same shenanigans with

[*] I find it also helps if you scream a little as you jump.

me? Under such circumstances, the metaphor of people as intuitive scientists is misguided. The truth matters, but the risk of getting it wrong in certain ways matters even more.

Phil Tetlock and his colleagues pointed out that in cases such as these, the metaphor of people as *intuitive prosecutors* might be more apt. Along with people's goal to understand they are also motivated to punish if they think others might be up to no good. In such cases, punishment is a form of self-protection much like jumping quickly when you think you see a snake. The more snakes you think are in your area, the quicker you jump in response to possible snakiness. The more disreputable sorts you think are in your area, the more likely you are to punish possible perps.

To test this possibility, Tetlock and his colleagues presented people with a brief video depicting a brutal assault and told them that the perpetrator either was or was not punished for his actions. Tetlock then asked the participants to judge a series of unrelated legal cases to determine whether the accused party had been negligent. When participants had been told that the brutal assault went unpunished, they were more likely to judge subsequent defendants as guilty of negligence than when they had learned that the perpetrator was duly punished. These findings suggest that people will take the role of an intuitive prosecutor rather than an intuitive scientist if they are worried that misbehavior might go unpunished. Tetlock also found that the prosecutorial mindset was more common among Conservatives than Liberals.

These data show us that generalized trust, and subsequent feelings of connection, play an important role in the different attitudes of people on the Right or the Left. These attitudes, in turn, interact with a variety of other beliefs to determine how people respond to others in need. For example, one of the largest differences between Left and Right lies in their reactions to harms that emerge from one's own behavior. As we discussed in the last chapter, Conservatives see people as largely responsible for their own outcomes, whereas Liberals see a variety of structural causes for success and failure. As a consequence, Conservatives are particularly punitive when they encounter self-induced harms; if you misbehave, you should suffer the consequences so you don't do it again. Conservatives' emphasis on autonomy in self and others leads them to feel less compassion when people suffer in ways they could have predicted and avoided. Under such circumstances, Conservatives are unlikely to feel the sort of connection that would induce them to help victims of misfortune.

Because Liberals see many external causes for misbehavior, such as poverty, prejudice, and discrimination, they are less focused on whether the harms people suffer are self-induced. Indeed, because the liberal worldview includes so many external causes for people's behavior, the concept of *self-induced harm* doesn't really make sense for the Left—the self is also a victim in this worldview. For example, if someone comes from a disadvantaged background, is it fair to say they caused their own problems or might their criminal record be a product of desperation or necessity? This worldview leads Liberals to feel a stronger connection to victims of misfortune, even if people's misfortunes are foreseeably brought about by their own behavior.

This differential focus on whether people are responsible for their own outcomes might seem like a trivial distinction, and sometimes it is, but it can have massive consequences. Consider the HIV/AIDS epidemic as a case in point. Most of the people who caught HIV/AIDS in the early years of the disease did so from unprotected sex, particularly unprotected homosexual sex among men. Many people have argued that it was the self-induced aspects of the disease that caused funding to lag so far behind the death rate in the first dozen years of the epidemic. Consistent with that possibility, the first major federal program to support people with HIV was named after a hemophiliac who contracted HIV/AIDS through a contaminated blood transfusion at thirteen years of age, despite the relative rarity of such cases.* Politicians realized that it was high time they did something to fight the epidemic, but they wanted to name the program in a manner that would lead everyone to feel a connection to AIDS victims.

There's More to Politics Than Left and Right . . .

In this chapter I've painted a picture in which people on the Left are focused on connection and people on the Right are focused on autonomy. But as we've discussed elsewhere, humanity can't be divided into just two types, so it's no surprise that not everyone fits neatly into a Left/Right category. What are these other folks like? Sometimes their politics are a smattering of different viewpoints, sometimes they're single-issue voters, but sometimes people aren't clearly Left or Right because they follow different political ideologies. Perhaps the best example of this latter group is Libertarians, who represent one in eight Americans.

Libertarians put liberty—individual freedom and autonomy—above all else. Neither social justice concerns nor respect for current norms are of sufficient importance to cause a Libertarian to sacrifice individual autonomy. Furthermore, because autonomy is mission critical for Libertarians, they deeply believe in the right of all individuals to choose their own fate. This belief leads to a key distinction between Libertarians and other people who are high in autonomy. If I happen to be low in connection and high in autonomy, I might value my own autonomy but I don't necessarily see the autonomy of others as more important than various competing values. In contrast, if I'm a Libertarian, not only do I value my own autonomy, but I also see the autonomy of others as more important than all other competing concerns. Autonomy

---

* His bravery in the face of institutional and personal bullying were undoubtedly a factor in the decision to name the program after him.

in self and others is not just a trump card, it's a sacred value for Libertarians.

What are the consequences of valuing autonomy above all else? First and foremost, the centrality of autonomy for Libertarians leads them to trust others to know their own preferences and decide their own fate better than anyone else can. That might not sound very radical, but if you walk this idea through to its logical conclusion, you see that it leads to some pretty wild policy implications:

1. Adults who want to sell their own body parts should be allowed to do so.

2. Adults who want to have sex with their siblings should be allowed to do so (perhaps with the proviso that they should use birth control to avoid poor health in their offspring).

3. Adults who want to kill themselves for any reason should be allowed to do so.

Do Libertarians really endorse such radical ideas that can be derived from the high value they place on autonomy? To test this possibility, Philip Tetlock and his colleagues asked people who followed a wide variety of political philosophies to take a survey. Tetlock asked these people to imagine they were lawmakers who were deciding if they would allow others to engage in various economic transactions. One set of transactions was meant to be routine—such as paying someone to clean your house or provide medical care—and another set was meant to be taboo—buying things that shouldn't be for sale, like a living person's body parts, votes in an election, or (my personal favorite) paying someone to serve your jail time. Participants were asked whether such transactions should be allowed or banned.

People's responses to the routine transactions looked pretty similar across the various political ideologies, but their responses to the taboo transactions broke into two camps: Libertarians and everyone else. Socialists, Liberals, and Conservatives were in complete agreement that the taboo transactions should be banned but Libertarians were happy to allow every one of them. What would it be like to live in such a world, where autonomy is sacred and held above all other values? First and foremost, it would be even nicer to be rich than it currently is. If Bill Gates were convicted of a DUI, he could open Fiverr on his phone as he was escorted out of the courtroom and offer $100,000 for someone else to give up their license and go to jail on his behalf. A gig that lucrative would be grabbed before he hit the sidewalk. If Elon Musk were convicted of murder, someone with terminal cancer and a family to support would assuredly accept a few million dollars to serve a life sentence or even the death penalty on his behalf. I'm not sure that justice would be served, but I suspect both buyer and seller would think they got a good deal.

This willingness to allow people to decide their own fate regardless of the consequences is a striking feature of Libertarianism. To get a sense of the psychology that underlies it, we need a more complete picture of Libertarian values. There's much we don't know, but my favorite portrait of Libertarians and how they compare to the Left and Right, is from a paper by Ravi Iyer and his colleagues. In their survey of Liberals, Conservatives, and Libertarians, they asked people to indicate how important five key values are to them (harm, fairness, loyalty, authority, and sanctity). A few things stand out in Figure 7.5, which depicts people's answers to these questions (as well as their empathy scores).

First, as we discussed earlier, Liberals care much more about harm and fairness than the other three values, which goes along with their higher empathy scores. Second, Conservatives show equal concern for all five values. Third, and perhaps surprisingly, Libertarians resemble Liberals in the fact that they care about harm and fairness more than the other three values and their feelings about loyalty, authority, and sanctity are a near-perfect match with Liberals. But Libertarians don't care about harm and fairness as much as Liberals do and they're not as empathic as Liberals or Conservatives.

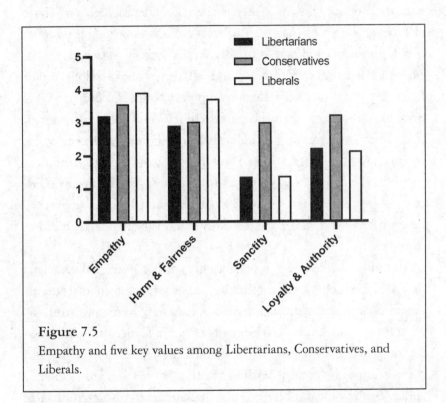

Figure 7.5
Empathy and five key values among Libertarians, Conservatives, and Liberals.

Due to their particular constellation of values, Libertarians are fiscally conservative, as big government and taxation are a direct threat to freedom and autonomy, but they're also socially liberal, as they view the behaviors of consenting adults to be no one else's business. They might disagree with what consenting adults do, but those disagreements remain secondary to their core belief that people should be allowed to do what they want.

What happens when Libertarians' values conflict with the values of the available candidates, as they inevitably must in a US political system in which one major party is (usually) fiscally conservative and the other is socially liberal? That is, who do Libertarians vote for if they don't pick a Libertarian candidate and must choose between their competing preferences? Given their strong emphasis on autonomy over connection, it shouldn't surprise you that they vote for Republicans more often than Democrats. But the share of Libertarian votes that goes to Republicans depends on whether the Republican candidate is fiscally versus socially conservative. For example, when the pragmatic and relatively non-religious senior Bush ran for president, he carried 74 percent of the Libertarian vote. Sixteen years later, when the more religious and socially conservative junior Bush ran for president, he carried only 59 percent of the Libertarian vote. When Ross Perot ran as a third-party candidate in 1992, the Libertarian vote was split almost equally across Clinton (32 percent), Bush senior (35 percent), and Perot (33 percent). Libertarians don't find themselves to be an easy match with either major party.

Because they're not a good match to either party, Libertarians provide an excellent test of our hypothesis that an increased emphasis on autonomy must, *by necessity*, lead to a decreased emphasis on connection. After all, it's possible there's something about being on the Left or Right that coincidentally causes autonomy and connection to be at odds with each other in people's political attitudes. But if we can find evidence that Libertarians de-emphasize connection, such a finding would suggest that independent of Left versus Right, people who are devoted to autonomy sacrifice connection. There are a few lines of evidence consistent with this possibility.

First, as you might recall, one of the key emotions that motivates connection is empathy, which is part of the reason why Conservatives and men are less focused on connection than Liberals and women. As you might expect, the percentage of males among Libertarians is even higher than it is among Conservatives. Furthermore, as we just saw in Figure 7.5, Libertarians are less empathic than either Conservatives or Liberals. Presumably these lower empathy levels allow Libertarians to consider Tetlock's taboo transactions without getting too worked up. You might also recall that collectivist societies are more focused on connection and individualist societies are more focused on autonomy. It should thus come as no surprise that Libertarians are more individualist and less collectivist than either Conservatives or Liberals.

These findings are consistent with the possibility that Libertarians resolve the inevitable trade-off between autonomy and connection by putting all their eggs in the autonomy basket, but it would be nice if we had more data on this point. We also don't know if Libertarians' strong emphasis on autonomy leads them to de-emphasize connection or if the low value they place on connection leads them to seek autonomy (or perhaps a bit of both). Whatever the exact order, the data with Libertarians provide further support for the idea that the trade-off between autonomy and connection plays a major role in people's political attitudes.

The findings in this chapter suggest that many political attitudes emerge from the emphasis people place on autonomy versus connection. On the Libertarian end of the spectrum, autonomy is paramount, meaning that pretty much any transaction between consenting adults is acceptable: people should be able to sell their votes or kidneys, they should be able to hire others to serve their jail time, etc. If that leads the rich to misbehave in some rather extraordinary ways, presumably the enhanced employment opportunities among the poor (who can now make extra cash by selling their votes and kidneys, and by serving jail time for rich lawbreakers) will compensate for the resultant societal upheaval.

When we move from Libertarians to Conservatives, autonomy remains more important than connection, but it's no longer sacred. Other moral values can trump autonomy for Conservatives, causing them to disagree with their Libertarian cousins when it comes to selling organs, votes, and jail time. Nonetheless, they're on the same page as Libertarians when it comes to governmental over-reach and taxation. And finally, over on the Left, we have Liberals who put connection above autonomy, and will line up to pay more taxes if that means increased spending on social programs that help disadvantaged people.

From this perspective, the enormous political divide in the US remains disheartening but not intractable. Disheartening because values that everyone endorses—but to differing degrees—lead to endless culture wars, failure to compromise, and interminable legislative paralysis. Tractable because it turns out that Americans agree with each other more than they disagree. Members of both major parties want the same things but prioritize their wants differently, suggesting that politicians who are skilled at bringing people together will always undo the harms caused by politicians who succeed only by demonizing the other side.

As we've seen in Part II of this book, autonomy and connection are deeply entwined with our identities. Men, individualists, Protestants, and Conservatives all emphasize autonomy while women, collectivists, Catholics, Jews, and Liberals all emphasize connection. Of course, people are members of more than one of these categories, so sometimes their identities will clash, but sometimes they'll work together. So far so good—there's no single best approach to life. But as we started to see in Chapter 5, and as we'll see throughout Part III, most people are now moving toward autonomy and away from connection. In the remainder of this book, we'll explore how and why this is happening and with what consequence.

Part III

Off-Balance

For the first time in the quarter million years we've walked this planet, our modern world has made our sociality unnecessary for survival. Our evolved psychology has not caught up with that fact, however, and the rush of freedom we feel from our extraordinary autonomy has left us unfulfilled, disconnected, and wondering whether we're missing out on something. In the next three chapters, I discuss the primary ways that our lives have shifted to emphasize autonomy over connection and how these changes have disrupted the balance between these two all-important needs.

8

Cities and the Great Shift to Autonomy

One of the most common questions adults ask small children is what they want to be when they grow up. I remember my five-year-old son declaring he was going to be an astronaut or a check-out guy at the grocery store. Although his interest in these careers faded with his love of pushing buttons, we can thank cities for the fact that anyone asks children this question in the first place. To ask our distant ancestors about their career preferences would have been bizarre; why ponder the issue when you have no choice. Everyone hunted and gathered, and that pretty much covered the available options. The advent of farming twelve thousand years ago meant there was now one more career possibility, but no one chose to farm over hunting and gathering. If you grew up in a farming community you became a farmer and otherwise you were a hunter-gatherer.

This state of affairs lasted for about five thousand years, until the rise of the first towns and then cities. Because farming is more successful on easily watered and fertile soil, farms initially sprouted up near riverbanks where water was plentiful and seasonal flooding created rich, alluvial plains. These farms not only supported the farmers who lived on them, they also created enough surplus to feed people who lived nearby and provided services to the farmers.

Initially those service providers gathered in small hamlets that dotted rich farming areas, but eventually these little villages grew until they consolidated into towns and then cities. The world's first city may well be Uruk, which sat on the banks of the Euphrates in what is now Eastern Iraq, south of Baghdad.

To get a sense of how humanity first began the shift from connection to autonomy, let's take a brief detour into the history of Uruk. As you'll see, the move to cities and an autonomous world demanded enormous cultural change, which emerged slowly over hundreds of generations before accelerating dramatically in the last century.

Uruk

The Euphrates has wobbled a fair bit in the last few thousand years and it no longer passes the remains of ancient Uruk. If you were to visit the archaeological site today, you might come away with the impression that Uruk was nothing more than a desert outpost, but in its glory it sat at the center of a massive farming community. Five thousand years ago, Uruk had a population of fifty thousand people, enormous temples, city walls, and was twice the size that Aristotle's Athens would become three thousand years later. Uruk was grand enough and old enough that it has the distinction of being mentioned in the Book of Genesis (10:10) as part of the kingdom of Nimrod (great-grandson of Noah himself).

Not coincidentally, Uruk is where writing was invented. Spoken language probably evolved as people struggled to communicate ideas that were separated from the speaker by time or space, such as what they were going to do tomorrow or how to get to a water hole many miles away. Such information was not as easily pantomimed as "Thag, there's a saber-toothed tiger behind you," and hence was more readily communicated with the help of vocabulary and grammar. We don't know when our ancestors gained the capacity for spoken language, but my guess is that it is older than our species, as *Homo erectus* were already planning and thinking about the future over a million years ago.[*] Because they could think about the future, *Homo erectus* would have benefited from the capacity to communicate about it as well, raising the possibility that they evolved rudimentary speech.

Writing, on the other hand, would have been of very little use to our hunter-gatherer ancestors. The primary purpose of early writing was to maintain an accurate record over time, particularly regarding details about which people might be inclined to disagree. Our hunter-gatherer ancestors told and retold stories around the firelight, where the details were often unimportant. Furthermore, with so many people aware of the stories that were being told—and everyone equally motivated to get the details right—there would have been many people to correct the storyteller if he misplaced where the hippo attack took place or where the water hole was located.

In contrast, if I've sold you a wagonload of wheat in return for the promise of six piglets next spring, we need some way to ensure that it's six pigs you'll deliver in early May and not five in late July. Because my memory might be inclined toward seven or eight piglets, but you might be inclined to recall that it was only four or five, we needed some way to record agreements to ensure that the details didn't change over time. Writing solved that problem by indicating what was stored where, who owed what to whom, and a long list of other economic transactions. Of course, like the wheel, the iPhone, and Silly Putty, once people invented writing they found lots of other cool uses for it.

One of the most important archaeological finds from Uruk is a list of occupations, apparently in order of importance. It might

---

[*] For example, *Homo erectus* carried their stone tools with them over great distances, suggesting they could envision a future in which those tools would be useful again.

seem like enormous good luck that such a list was found, but it is one of the most common writing examples found in ancient Mesopotamia, suggesting that scribes practiced writing it to develop or demonstrate their skill. Thus, at the very least, we know from the outset that there were scribes in ancient Uruk.

The list of occupations hasn't been fully translated, due to the many challenges of interpreting a form of writing in a language that no longer exists, but we've translated enough of it to know that people in Uruk did lots of different things for a living. At the top of the list, there were numerous types of leaders, including city leaders, assembly leaders, law leaders, military leaders, and even plow and barley leaders. Down a notch or two from these various leaders, we have advisors and ambassadors.* Further down the list we get to varieties of priests and eventually we find people who actually work for a living, including stonecutters, smiths, shepherds, gardeners, cooks, bakers, potters, weavers, jewelers, and the like. There's no mention of jugglers, but otherwise the list of occupations looks a lot like what you'd see at your local medieval fair.

These jobs all required a fair bit of expertise, but there was also an enormous need for unskilled labor. For example, people were employed to carry things around the city and between the city and the countryside. Initially this was done almost entirely on foot by porters, but soon donkeys were incorporated as pack animals, river boats were used to move goods, and eventually wheeled wagons were used as well. We know little about the life of these laborers, but they appear to have been paid a common daily wage in the form of food, perhaps from a central city authority. How do we know this? Thousands and thousands of low-quality, mass-produced food bowls have been dug up from sites around Uruk. The uniformity and ubiquity of these bowls suggest they were a standard way for

---

* Suggesting that entourages were invented within moments of people accumulating enough wealth and power that it was worth sucking up to them.

people to receive their daily wage, which in turn suggests that the thousands of people who used these bowls were all engaged in jobs that paid about the same. Many of these laborers were probably free to do whatever work they chose, but we also know that the occupational ladder was anchored at the bottom by slaves who had been captured in war, convicted of crimes, or were so heavily indebted that they were forced into servitude.

As is evident from this discussion, there's a lot of inference involved when we have so little raw data to guide us. It's possible, for example, that these food bowls were used by individual employers who bought them at the Uruk equivalent of Walmart and used them to pay their workers from their private stocks rather than from some centralized location in the city. But we also know that in the early days of Uruk, people were buried in a uniform fashion, with no distinction between those who might have been richer or poorer. These uniform burials suggest that community allegiance was paramount, in that people were buried as citizens of Uruk rather than as leaders, priests, or masons. By the time Uruk was at the height of its power, burials were differentiated by wealth, with some people buried with valuables such as copper ornaments and others simply placed in the clay without adornment. These early uniform burial practices raise the possibility that labor might have been organized by city authorities, rather than by various individuals doing their own thing.

In this spirit of inference and trying to go beyond the data given, let's consider their reliance on wool for clothing. We know from various archaeological finds that wool was spun into textiles to make clothes, as people in Uruk no longer wore animal skins that were common among hunter-gatherers. We also know that wool takes on dye very easily and can be colored in complex ways. Colors could have been used as a fashion statement, as they are today, but they might have been used to indicate occupations. In today's world we design uniforms to signify occupations that are important to the general public, such as the police and military, and we also have uniforms that emerge as a by-product of job demands. For example, cowboys wear spurs and cooks wear aprons. These occupational signifiers are handy when you meet such people, as you know who to ask for advice about horseback riding and who would know if the apple crumble is gluten-free.

I suspect that in ancient Uruk these occupational signifiers were much more than a convenience. When cities were new and people were moving to them for the first time, it must have been intimidating to encounter so many strangers. Living among strangers might not seem odd anymore, but it was a seismic shift for our ancestors, who almost never encountered strangers and for whom such encounters were fraught with danger. We now have rules for how to interact with people we don't know (don't stare at them, use polite forms of address when you talk to them, don't ask personal questions) and we give the situation very little thought, but such cultural rules probably took generations to develop.

One way for the first city dwellers to make the experience more manageable would have been to ensure that everyone knew everyone's occupation, again because that information helps you know how to interact with others and what their status is in the community. It thus seems possible that everyone might have worn color patterns that signified occupations to help others know who they were. The historical record suggests that occupations in the early days of cities were clan-based, with certain families filling certain jobs, as they often do in small-scale societies today. If that's the case, such occupational uniforms would also provide information about the family background of the wearer. That knowledge, in turn, would have reduced the feelings of anonymity and trepidation that the first city dwellers felt as they started encountering so many people they didn't know. Again, this is all conjecture based on what we know about wool, dyes, and occupational signifiers in agrarian societies today, but it's possible that everyone in Uruk was identified to others by their occupation and hence by clan.

Regardless of the details, what we know of life in Uruk five thousand years ago paints a picture in which autonomy was still second fiddle to connection. For the first time in human history, there was a wide range of occupations—well over a hundred from what we can tell—but most people didn't choose them any more than they chose whether to hunt, gather, or farm. Rather, if you're born into a clan of bakers, baking is what you did. From that perspective, connection and the obligations and responsibilities associated with it were still paramount. Nonetheless, by creating the demand and opportunity for new jobs, cities opened the door to the sort of choice we have today, in which my five-year-old son contemplated whether he'd rather operate a rocket ship or a cash register.

Over time, probably measured in hundreds of years rather than individual life spans, this system of caste- or clan-based occupation slowly loosened to allow people to pursue their own proclivities. Steve's dad is a baker but Steve is big, burly, and allergic to flour? No problem. Steve can apprentice out to Rick's dad to learn how to blacksmith, while Rick pursues his interest in the arts with Tom's dad. . . . Once such a system got underway, in which people were allowed to choose what they wanted to do for a living, products and services would have improved by matching people to occupations based on their preferences and capabilities. More importantly, career choice represented the first major increase in autonomy since *Homo sapiens* first walked the planet over a quarter million years ago. We owe this enormous change to cities and the specialization they enabled and rewarded.

There's no such thing as a free lunch and the autonomy enabled by cities is no exception to this rule. The increase in autonomy that cities gave us came with costs, the most notable of which is the inevitable price that had to be paid in connection. Cities enabled specialization but they also pulled us out of the close confines of family and friends. Once we moved to cities, we no longer spent our entire lives with a small group of people we knew incredibly well. There are benefits to that change, as it's nice to have lots of choice when it comes to friends and romantic partners. It's also nice when you don't feel like everyone's in your business (a point I return to later in this chapter). Nonetheless, the sense of deep connection that you feel when everyone knows you well is lost once you start interleaving your life with strangers.

Town and Country, Then and Now

To state the glaringly obvious, the world has changed a lot since the first cities got started. To state the less obvious, the pace of that change has accelerated in the last few hundred years and then again in the last few generations. Although cities are over five thousand years old, they didn't really catch on until two hundred years ago. Prior to 1840, when populations began moving to cities en masse,* fewer than one in ten people worldwide lived in a city. By 1960, this number was one in three. After thousands of years of very slow growth, in 120 years the percentage of people living in cities tripled.

Rather than slowing down, the pace of urbanization has only accelerated since the 1960s. In the last sixty years, the ratio of people living in the country versus the city has flipped, with people in cities outnumbering those in the country for the first time in human history in 2007. Planet Earth used to be a rural place but it's not anymore. We now have only three-quarters of a person in the country for every city dweller, and by 2050 we'll have less than half a person in the country for every city dweller: an exact reversal

---

* More specifically, Americans began flocking to the cities in 1840. It was closer to 1900 before the rest of the world began shifting to cities as well.

of where we were in 1960. This fundamental change in the way we live is so unprecedented and rapid that our psychology simply hasn't kept pace, with notable costs to our life satisfaction (a point I return to in Chapter 11).

Despite these large-scale and ongoing changes in human living conditions, there's still a lot of folks in the country—3.4 billion to be precise. That's more than the population of the entire planet in 1960, so there are still plenty of places where everyone has known everyone else most of their lives. We can rely on these places to compare people's connection and autonomy when they live in a world of closely known others versus a world of strangers. Furthermore, we can make these comparisons in highly mobile and industrialized societies, like those in North America and Europe, but also in less mobile and more agrarian societies, like many in South Asia, Africa, and South America. Country folks in highly mobile societies encounter a fair few people they don't know, but country folks in agrarian societies are surrounded almost entirely by people they've known their whole lives. As a result, comparisons between city and country and between industrialized and agrarian societies allow us to test the impact of cities on autonomy while also giving us a snapshot of the attitudes and opinions people held in a world in which everyone knew everyone else.

If city living enhances autonomy and reduces connection, and if highly mobile industrialized societies also enhance autonomy and reduce connection, then the differences between city and country should also emerge between industrialized societies and more traditional ones. Compared to people who live in rural areas, people in cities should emphasize autonomy more and connection less, regardless of what society they live in. Similarly, compared to people who live in traditional societies, people in industrialized societies should also emphasize autonomy more and connection less, regardless of whether they live in the city or the countryside.

To test these possibilities, we can analyze data from the World

Values Survey.* The latest wave of the World Values Survey includes people from sixty-four different countries who live in large cities all the way down to small villages and hamlets. To provide the clearest contrast between people who live among strangers and people who live entirely with others they know well, I compared the opinions of people who live in capital cities to the opinions of people who live in the smallest rural areas. I then picked a list of eight industrialized nations† and another list of eleven largely agrarian nations‡ as my set of countries to investigate. Finally, I considered only native-born citizens of the different countries I examined. These decisions gave me a sample of slightly over ten thousand people, scattered across the small towns and largest cities of nineteen different countries.

The World Values Survey contains a few questions that allow us to test the hypothesis that urban living and industrialized societies emphasize autonomy while rural living and traditional societies emphasize connection. For example, in one set of questions, people are asked the following: "Here is a list of qualities that children can be encouraged to learn at home. Which, if any, do you consider to be especially important?" They are then presented with a list of eleven traits, such as "good manners" and "hard work." Two of these traits provide a test of our hypothesis—*obedient* and *independent*.

Obedience and independence are both valuable traits in children, but parents who live in societies that emphasize autonomy should believe it's particularly important for their children to be

---

* Another remarkable, publicly available dataset that contains the largest survey of people's beliefs and values around the world: see https://www.worldvaluessurvey.org /wvs.jsp.

† Australia, Canada, Germany, Great Britain, Greece, Netherlands, New Zealand, and the United States.

‡ Bolivia, Ecuador, Ethiopia, Guatemala, Kenya, Myanmar, Nigeria, Peru, Uruguay, Venezuela, and Vietnam.

independent, as independence is the clearest expression of autonomy. In contrast, parents who live in societies that emphasize connection should believe it's particularly important for their children to be obedient, as roles and responsibilities (and hence obedience)

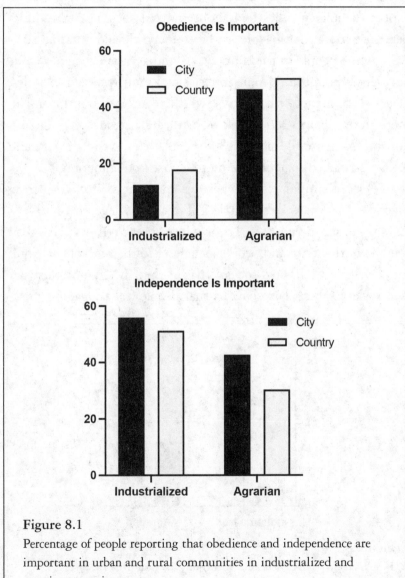

Figure 8.1
Percentage of people reporting that obedience and independence are important in urban and rural communities in industrialized and agrarian countries.

are critical in societies that value connection. To test these possibilities, we can examine the percentage of parents in urban and rural areas in industrialized and agrarian countries who indicate that each trait is especially important.

As you can see in Figure 8.1, people's opinions line up as we expect. People in cities and industrialized societies value independence more than people in the countryside and agrarian societies, while people in the countryside and agrarian societies value obedience more than people in cities and industrialized societies. These data suggest that people who live in the country and in more traditional societies seek to inculcate a strong sense of connection in their children, whereas people who live in the city and in industrialized societies are more focused on autonomy.

We see a similar effect when we move beyond child-rearing practices to ask how close people feel to their village, town, or city. As you can see in Figure 8.2, people feel closer to their village than they do to their city and people feel closer to their village or city if they live in a traditional society than if they live in an industrialized society. These data show us that people feel a greater connec-

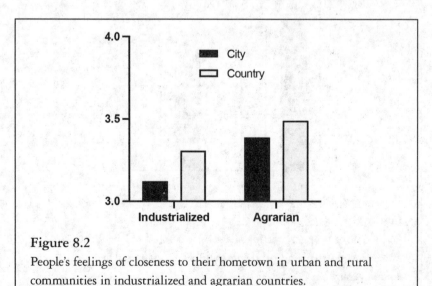

Figure 8.2
People's feelings of closeness to their hometown in urban and rural communities in industrialized and agrarian countries.

tion to the people around them when they live in small villages and traditional societies than when they live in large cities and industrialized societies. Connection is key when you spend your entire life among the same group of people.

These opinions of people around the world are consistent with the idea that urban life emphasizes autonomy and rural life emphasizes connection, but they don't tell us much about people's actual behavior. Although we'd expect attitudes and behaviors to correspond, it's possible that various factors push them in different directions. For example, perhaps it's easier to meet your neighbors when you have hundreds of them in your apartment building than when you only have a few of them a dozen miles down the road. If that's true, city folks might care about autonomy more than country folks but still experience more connection because they live cheek by jowl with so many other people.

The data suggest otherwise. Rural Americans are more likely than urban or suburban Americans to know their neighbors. Furthermore, when they were asked in a 2018 Pew poll whether they have a neighbor they would trust with their house keys, rural residents were more likely to say yes than suburban or urban residents. Despite the distances between neighbors in rural areas, country folks are more likely than city folks to know and trust their neighbors, with responses in the suburbs falling somewhere between the two.

These differences between life in the cities and life in the countryside didn't emerge overnight, but each generation of city dwellers has drifted farther and farther from their neighbors. Not long ago, people knew everyone who lived around them, as their families had comingled for generations. But residential mobility and isolation have both increased over time, with the result that neighbors have become less neighborly. In the early 1970s, 30 percent of Americans spent time with their neighbors at least twice a week and only 20 percent never spent time with their neighbors at all.

Fifty years later these figures have reversed, with over 30 percent of Americans never spending time with their neighbors and fewer than 20 percent spending time with them at least twice a week. Cities pack people in like sardines, but apparently sardines who barely speak with one another.

Feeling Different in a Small Town

The internet enabled all sorts of wonderful effects, the most important of which may be the democratization of information. Prior to the internet, if you wanted to look up a fact or read a classic text, you had to have money or access to a large library. After the internet, most people can look up any fact or read any classic—the world's knowledge is available to anyone with access to a browser.[*] If I had to identify a positive effect of the internet to put in second place, I'd choose the way it allows people who feel different to realize they're not so different after all. Prior to the internet, if something about you was sufficiently unusual that you were one in a hundred or one in a thousand, you felt isolated and alone. This effect was particularly strong if what made you unique was also socially undesirable, such as having a non-normative sexual orientation or a locally unacceptable religion. People with such rare qualities quickly learn to keep their differences to themselves, as others are not always accepting.

Once the internet took off, however, no matter how unique you were, you could find others who shared your traits. You're the only Zoroastrian in your school? No worries, an online fire temple is only a click away. You think the Earth is flat but your friends make fun of you when you raise the issue? No worries, everyone else in the Flat Earth Society feels the same way you do. It doesn't mat-

---

[*] Two-thirds of the world's population, as of this writing.

ter that Flat Earthers are incredibly rare and scattered around the globe;* you can still meet them online.

And this leads to a major downside of living in the country. The main costs of connection we've considered so far are the constant demands that emerge from your roles and responsibilities. But there is another cost to the type of connection that exists in small communities: the cost of being different when different equals weird. One of the key factors determining whether a culture is individualist or collectivist is the homogeneity of the populace. When there is a great deal of variability in ethnicity, religion, etc., cultures tend to be individualist, as the overriding ingroup identity must compete with a wide variety of sub-identities. Homogeneity, in contrast, promotes interdependence, collectivism, and pressures to conformity. By virtue of their smaller size and historical origins, country towns tend to be much more homogeneous than cities.

Cities are to the countryside what the internet is to society. Once you move to a city it no longer really matters if you're one in a million, as plenty of other people are just as unique as you are. But in the country it matters a lot; if you're a rare individual you can count on being the only one. We talk a lot about the value of diversity, and for good reason: diversity of opinion, background, and approach can be valuable as people try to devise solutions to the problems they face in business, science, or life in general. But like everything else, diversity comes with costs, and those costs are most evident when diverse members of a population are also solo representatives of their group. When you're the only Jew in a group of Christians, the only Republican in a group of Democrats, or the only Black guy in a group of White people, everyone looks to you as a representative of your group ("I'm curious, Bill, what do Jews think of Trump?"). Although we know there's a range of opinions

---

* Heh heh.

within every group, somehow we forget that fact when groups are represented by only one or two individuals. As a result, solo minorities tend to feel stereotyped and put on the spot, which is not exactly conducive to a free-flowing exchange of ideas.

In the country, people who have rare traits or identities are almost always the only person like themselves for the simple statistical reason that their attributes are rare. Because their rare characteristics can be socially unacceptable among people who don't share them, such people often hide their rare qualities, leading their attributes to seem even rarer still. As a consequence, diversity is rarely celebrated in country towns. Rather, people who grow up in small towns looking or thinking differently from everyone else often can't show their true colors for fear of rejection. For these folks, the strong connections that give everyone else comfort feel like a cruel version of hide-and-seek in which they risk being exposed and humiliated. Those are not the sort of connections that make us feel safe and secure, nor do we look back on them with nostalgia.[*]

Country Folks and Self-Reliance

Cities made it possible to choose any career, massively expanding autonomy. Because one of the most important aspects of autonomy is self-reliance, you might think city folks should be more self-reliant than country folks. If this argument doesn't sit right, it's because it shouldn't. An ironic consequence of urban/rural differences in specialization is that self-reliance is more important in the country. In the city, no matter how esoteric your need, you can find someone who solves that need for a living. You need the insulated padding on the edge of your refrigerator door replaced? No

---

[*] If you're interested in reading further on this topic, I recommend this personal account: https://www.washingtonpost.com/opinions/2023/07/24/small-towns-racism-ohio-jason-aldean.

problem—there are six refrigerator insulation companies to choose from. Your carbon fiber bicycle isn't changing gears like it used to? No worries—there are bicycle repair companies who can fix your fancy derailleur in a jiffy.

The person who insulates your refrigerator door or repairs your bike might not be much use at anything else, but a lifetime dedicated to refrigerator or derailleur repair ensures that the specialist knows more about your problem than you could possibly learn on your own. Specialists like that don't live in the country, as they'd spend their day waiting for the phone to ring. If you live in the countryside and need something built or repaired, you'd better be able to do it yourself or with the help of a few buddies who've grown up to be self-reliant just like you. The absence of specialists in the countryside means that everyone needs to be a human version of a Swiss Army knife.*

We can see this rural/urban difference in self-reliance in a variety of ways. First and foremost, country people drive pickups, which are the ultimate do-anything vehicle. No matter the project, it'll be easier with a pickup—and the bigger the better. An F150 might be tricky to park at your favorite record shop in the East Village, and it might not squeeze past the people who've double- and triple-parked on your street while they run a quick errand, but those aren't problems in the country. Pickup trucks are the most popular vehicle sold in thirty-nine of the fifty states, every one of which has a substantial rural population. I've lived in a city most of my adult life, but I still get nostalgic for my hometown when a big pickup rumbles by.

Bruce Springsteen sings more eloquently about cars in urban

---

* I'm not sure what the opposite of a Swiss Army knife is—a specialized tool that does only one thing, does that one thing very well, but is useless at everything else. Perhaps city folks like myself are best characterized as grapefruit spoons? Or maybe we're bundt pans?

life than anyone else. He's got a '69 Chevy with a 396 ("Racing in the Street"), he's driving a stolen car on a pitch-black night ("Stolen Car"), and he dreams of his girlfriend's pink Cadillac* ("Pink Cadillac"), but he doesn't have much to say about pickups. Country musicians have lots to say about pickups, which play a prominent role in country music as they do in everyday life. People use their pickup trucks to meet up in Florida Georgia Line's reminiscences about small-town Friday nights ("Long Live") and in Brad Paisley's ode to partying before and after the show ("Out in the Parkin' Lot"). Meanwhile, Carrie Underwood reminds us that pickups are also a place for couples to get cozy with each other ("Out of That Truck"). As these and many other country songs make clear, pickups may enable self-reliance but they're also much better at helping people connect than a Honda Civic† ever could be.

It's not just pickups that reflect the importance of self-reliance in the country. Guns, hunting, and even fighting all play a prominent role in songs that resonate in the countryside, and all three are instantiations of self-reliance. Guns ensure you don't have to call possibly distant police to solve your security problems, even if they do add to other people's security headaches. Hunting ensures you don't need the local grocer to feed your family. And fighting is one of the oldest ways to solve your disagreements, even if it doesn't actually reveal who's right and who's wrong.

So what are we to make of these seemingly contradictory urban/rural differences in connection, autonomy, and self-reliance? I'd suggest the following: First, connection is strong in the country because everyone knows and depends on each other. Second, part of the reason humans evolved the need to connect is due to the increased effectiveness of cooperative groups compared to single individuals. Thus, it's no surprise that the connections people form

---

* Yes, that's what he means.

† The most popular car in California.

in the country don't make anyone feel less capable of dealing with a demanding environment, nor do they diminish the importance people place on that effectiveness. If anything, strong connections enhance the importance of self-reliance, as everyone is expected to look out for themselves, their family, and their friends.

From this perspective, it's clear that greater self-reliance in the country isn't a sign that people want to strike out on their own any more than self-reliance among hunter-gatherers was a sign that they were seeking independence and autonomy. Rather, self-reliance is a manifestation of the need to solve the various problems people face in the country. By way of example, consider the combination of self-reliance and cooperation that can be seen in the barn-raisings of the Amish, when they come together to put up a new barn for one of their members. Everyone is expected to pitch in, everyone is expected to know how to do the numerous tasks that are involved, and the fact that everyone does their part ensures that an incredibly demanding job can be knocked off in a single day.

Country people typically don't work together as entire communities like the Amish do, but they do call upon friends and family when faced with a job that's too much for one individual to handle. The essence of self-reliance is knowing how to do things so you can take care of yourself, but people also develop self-reliance in the country to be of value to others. Country folks can't call upon the expertise that's available in the city, but because they're all Swiss Army knives with tight connections to each other, they solve their problems anyway.

Moving to the City and Leaving Happiness Behind

When you consider the complete picture, country living has a range of pluses and minuses. The worldwide consensus is that the minuses outweigh the pluses, which can be seen in the fact that populations have been shifting from the countryside to cities for two hundred years. When people vote with their feet with such consistency, we need to take their preferences seriously.

But that doesn't mean the decision to move to the city is cost-free—far from it. The gains in autonomy have come at a substantial cost to connection. We've seen that cost in the loss of trust people feel for their neighbors and in the loss in connection people feel to their hometown. But we haven't yet delved into their feelings about these costs. Do people feel the price they're paying when they live in the city? Do the costs outweigh the benefits?

We can investigate these questions in two different ways. First, if the declines in trust and connection that we've discussed are meaningful, people in the city should be less satisfied with their friendships than people in the country. If you've got no one you can trust with your house keys, it stands to reason that your friendships have become less meaningful. To test this possibility, we can return to the General Social Survey to measure Americans' satisfaction with their friendships when they live in cities of over a quarter million versus country towns with fewer than twenty-five hundred people.

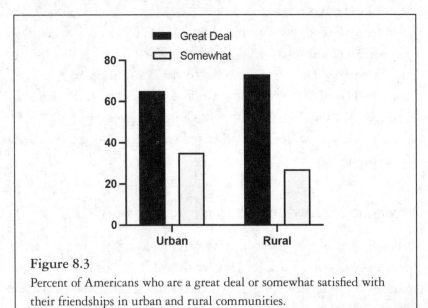

Figure 8.3
Percent of Americans who are a great deal or somewhat satisfied with their friendships in urban and rural communities.

As you can see in Figure 8.3, small-town Americans are more satisfied with their friendships than city dwellers. The good news is that hardly anyone is actively dissatisfied with their friendships, so this analysis is really comparing people who are very satisfied with people who are only somewhat satisfied. Nevertheless, these data show us that small-town folks are more likely than city folks to be very satisfied with their friends. The effect isn't very big, but it's stable across age groups and it emerges even though people have much less choice regarding their friendships when they live in the country than the city.

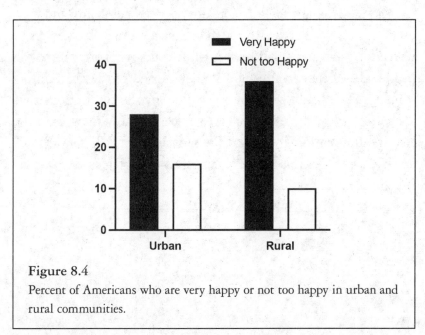

Figure 8.4
Percent of Americans who are very happy or not too happy in urban and rural communities.

Does satisfaction with friendships translate to happiness in general? The answer seems to be yes. Despite all the disadvantages of country living, and despite the fact that people are leaving the country in droves, people are happier in the country than they are in the city. As you can see in Figure 8.4, these differences aren't terribly large, but they're there, and again they're stable across age groups. People in the country are more likely to be very happy and less likely to be not too happy than people in the city. The size of this effect is almost identical to the differences we see in friendship satisfaction, in part because these two feelings are tightly related.

Cities may pull in people for the opportunities, but the warm feelings we lose when we're no longer part of a tight network of connections matter. Although rural/urban differences in happiness have lots of potential causes, they raise the possibility that our lives have shifted too far toward autonomy and away from connection. The pluses and minuses of city living are often conceived in terms of entertainment and opportunity versus chaos and cost, but the underlying trade-off is really between autonomy and connection. City dwellers yearn for the connection that is common in the countryside and small towns, while country folks imagine the independence they would gain if they moved to the city. Unfortunately, the pull of autonomy overwhelms the connections that anchor us, with the result that we've been steadily moving to the city and undermining our own life satisfaction. We'll consider additional aspects of this problem in the next chapter, as we examine the impact of education and wealth.

9

Education, Wealth, and Supercharged Autonomy

Cities started the world on the path toward specialization, but becoming a specialist wasn't originally based on a formal education. Rather, trades were a family occupation and children were expected to help family members, in part to pick up the relevant skills. This sort of learning-on-the-job apprenticeship is highly effective and still exists for many trades. The problem with apprenticeships, though, is that they don't teach you skills beyond what you'll need for your chosen occupation, so you have limited capacity to pivot if your career plans change. In other words, apprenticeships don't provide the sort of autonomy-expanding knowledge that forms the foundation of a broad education.

Growing up in the United States with parents who went to college, I always assumed that I would go to college to get a liberal arts degree (by which I mean a degree that requires students to take courses in a wide variety of fields). I knew full well that a liberal arts degree doesn't actually prepare you for a job,[*] but I also knew

* As indicated in the well-known joke. Q: "How do you get a recent psychology graduate off your porch?" A: "Pay for the pizza."

it would give me two things I wanted. First, it would provide me with the sort of broad background in the humanities, sciences, and arts that would make me a well-rounded person. Second, it would teach me a little bit about a lot of things, giving me exposure to enough topics that I would find a field that suited me.

When I moved to Australia twenty-some years ago, I discovered that a liberal arts degree was the exception rather than the rule. Most students here dive into their major in their first year of university and take only a few courses outside of it. My first reaction was that Australians were missing out on the opportunity to learn lots of interesting stuff. Over time, however, I've come to realize that the real cost of a focused education is not in lost tidbits that are handy when chatting with your Uber driver. Rather, the price of a focused education is paid in terms of autonomy. By encouraging students to choose their course of study straight out of high school, the Australian educational system leads many people to make ill-informed choices. As a result, what Australian students might have wanted to do and what they end up doing can be two entirely different things. That happens everywhere, but the advantage of a liberal arts degree is that you expose yourself to a wide array of ideas before you make a decision about what you'd like to be.

In my own case, I applied to college intending to become a physicist, but a single semester of vector calculus alerted me to the fact that I wasn't cut out for it. My next thought was that I would be a lawyer, but I was rapidly cured of that delusion by my inability to read a single paragraph of Gunther's *Constitutional Law* without my mind wandering.* After a bit more noodling around, I eventually settled on psychology. Although I still remember lots of stuff from my courses in political philosophy, ancient Greece, and more, the biggest gift my liberal arts education gave me was the chance to

---

* Now in its seventeenth edition, suggesting that other students found it more engaging than I did.

learn broadly enough to know I'd found a topic that suited me. Education gave me the power of genuine choice, which is the essence of autonomy.

Four years spent studying the liberal arts might give you a greater understanding of your career options, and hence more autonomy, than almost anything else, but we see effects of education at all levels. To start with the unsurprising, people who have a better education are more likely to get jobs they find fulfilling. Thus, one way education gives you autonomy is by enabling you to do work you want to do. The data showing that education leads to fulfilling work are clear and consistent, but we don't know if people who pursue more education would still pursue more rewarding jobs even if they didn't have a good education. Maybe their personality or initiative leads them to get a good education *and* a good job, and education is not really causing anything. Fortunately, there are experiments that address this possibility.

The most important such experiment in my mind is by the economist Raj Chetty and his colleagues, who followed more than ten thousand Tennessee kindergarten students into adulthood. These people were randomly assigned to different classrooms from kindergarten through grade three, thereby ensuring that it was nothing about them or their parents that caused them to enroll in different teachers' classrooms. Chetty and his colleagues found that children who were allotted smaller classes (~15 students), classes with more experienced teachers, and classes in which their classmates performed better, were more likely to go to college than children who were put in larger classes (~22 students), classes with less experienced teachers, and classes in which their classmates performed worse. Furthermore, students who were lucky to be assigned to high-performing classrooms also found more lucrative jobs by their late twenties. Because the students were randomly assigned to their classrooms, we can be confident that these effects are causal in nature; small classes with good teachers early in one's school experience increase the chances that students will go on to college and earn more money as adults.

One of the striking things about these findings is that the effects of such early-childhood interventions disappear by eighth grade if the outcome of interest is standardized test scores. So why do Chetty and colleagues find teacher and classroom effects on university attendance when there are no lasting teacher or classroom effects on test scores? And why do the effects Chetty documents last well into the students' late twenties? We don't have definitive answers to these questions, but Chetty and colleagues examined teachers' evaluations of students' behavior when they were in grades four and eight, where they found that children who had been in high-quality classrooms displayed more initiative, greater effort, and less disruption.

How does individual attention from a teacher, particularly a good one, give you initiative and self-discipline? I suspect that high-quality teaching helps children realize they have good prospects in school, leading them to develop a love of learning. Recall that the primary purpose of autonomy is to help you find success by dedicating your effort to domains in which you feel you have good prospects and withdrawing your effort from domains in which you don't. By showing children they can succeed in school, good teachers in small classes also show children that school is a domain in which it would be worthwhile to expend their best efforts.

We see similar effects with brief interventions that are targeted specifically at students' sense that they can do well in school. For example, in an experiment by Geoffrey Borman and his colleagues, all the students starting middle school in a Midwestern school district were assigned to either a *good-prospects* or a *control* writing task. In the good-prospects task, students were asked to write about their worries that they might not fit in or do well now that they're in middle school. They first read about how prior students had some of the same worries but overcame them and then wrote about how they, too, might overcome such concerns. In the control task they wrote about how they might develop an interest in politics or become accustomed to the noise level in the cafeteria now that they're in middle school. Similar to the results from the Tennessee kindergarten study, Borman and colleagues found that students who engaged in this brief good-prospects task had better grades at the end of the year, as well as fewer absences and fewer disciplinary problems. This brief intervention seemed to give students confidence that their worries were widely shared, that other students had overcome them, that they would too, and that they really did have good prospects in school.

These findings suggest that better early education leads children to *choose* to try harder in school, which then leads to autonomy-enhancing rewards later in life. It may seem remarkable that kindergarten can have such lasting effects, but it makes sense that by getting children excited about education just as they start their journey, good teachers in small classes have a lasting impact (in the same way that Ronny had a lasting impact on my own educational journey when he mentioned in first grade that I was the smartest kid in class). This possibility suggests that the positive effects of education on autonomy compound over time, due to their immediate impact on people's choices and the larger array of choices that subsequently become available.

But school does much more than prepare you for a good job. Important though careers are, the effects of education on autonomy go well beyond job opportunities. First and foremost, education opens your mind. In the absence of education, it's natural to assume that your own reality is universal. Unless you see counterexamples, it makes sense to extrapolate from your family to other families, from your village to other villages, and from your nation to other nations. My father died last year, and one of the most touching things about his funeral service was the number of our childhood friends who spoke about how important he was in their intellectual lives. I hadn't realized it, but it turns out he was instrumental in getting them excited by the world of ideas—for example, by showing them that families could discuss and debate at the dinner table rather than just watch TV.

In this particular case, my father's counterexample made a big difference to my friends, but you can find your own counterexamples if you can read. By giving you access to every culture, every historical event, and every major idea anyone has ever had, books expand your world like nothing else. If you don't know what's possible, if you don't know which rules are human universals and which are cultural idiosyncrasies, you'll spend your mental life locked in a powerful but invisible prison. Just as fish (presumably) don't know they're in water, people who can't read don't know when they're living under arbitrary rules rather than the rules they might prefer *if they were aware they had a choice.* Mind you, reading doesn't guarantee independence, but it does provide a giant step toward autonomy by making you aware of the alternatives.

My friends learned from my father that intellectual debate can be a healthy family activity, but the opportunity to learn from someone else's father is only available when the cultural rules are sufficiently loose. Particularly if you live in a small village where literacy rates are low, cultural rules tend to be hard and fast. If mothers in your village do X and fathers do Y, you won't see any fathers doing X, so there won't be any counterexamples of how things can be different. The importance of this aspect of literacy hit me about fifteen years ago, when I had the good fortune to spend my sabbatical as part of a multidisciplinary group at the Wissenschaftskolleg (Institute for Advanced Study) in Berlin. The person who charmed me the most during my time there was a historian from Senegal. Whenever possible, I sat down with him for lunch or dinner to hear about his life and research.

One day we were chatting idly about our families when he told me about a journey he and his family took that involved walking from his village to a larger town where they could arrange transport to Dakar. His wife was heavily pregnant at the time and the weather was incredibly hot, so he was carrying their firstborn child to ease the burden of travel on her. As they passed through a small village on their journey, people poured out of their homes to see who these strangers were and what they were doing. When they saw my friend carrying his child, they were shocked and dismayed, asking why his wife wasn't carrying their toddler. He pointed out that she was pregnant and didn't have the energy to carry their child in the hot sun on such a long journey. The villagers found his answer unacceptable and warned him that evil would befall him if he continued to carry his child. When he tried to explain that he does it all the time and that lots of fathers carry their children, they would have none of it, demanding that he give the toddler to his wife. When he stood his ground, the villagers insisted that he and his family leave immediately and circumnavigate the village rather than walk through it, lest he bring evil upon them.

There was no school in this tiny village and none of the villagers could read. Had they been literate, someone would have known that what he told them was true—that fathers often carry their children, and that the village was not going to be cursed if they allowed him to carry his child as he passed through. But because they couldn't read, they were afraid to bend their local rules, believing them to be universal and fundamental rather than idiosyncratic and arbitrary. Literacy, and the world of ideas to which literacy provides access, would have given them the power to decide such issues for themselves, increasing their autonomy in ways they didn't know they were missing.* Again, literacy doesn't give you the right answers—plenty of literate people have inane ideas—but it does give you the power to decide issues for yourself

* For what it's worth, countries with lower literacy rates also tend to have much stricter gender roles.

rather than blindly following local norms. That power is the essence of autonomy.

Because literacy is so important in the development of autonomy, some of the biggest effects of education can come from simply attending primary school. By way of example, consider the research of Endale Kebede and his colleagues. They were interested in fertility rates among Sub-Saharan women and they found a common negative relationship between years of schooling and fertility. In their study, women in Kenya and Nigeria who had completed primary school had three fewer children than women who had no formal education. Their analysis also revealed that the political turmoil that disrupted education in Africa in the 1980s resulted in the addition of about half a child per woman, suggesting a causal relationship between education and fertility.

You might wonder what these fertility data tell us about autonomy. After all, a person can *choose* to have a large family. But high levels of fertility also result from obligation (in cultures that expect women to have many children) or lack of knowledge, and hence lack of choice. Across the developing world, the data suggest that falling fertility rates are primarily a product of female empowerment, which itself is typically a by-product of increasing education and wealth. Once women gain access to affordable and reliable contraception, family size tends to shrink. These data implicate lack of contraception, and hence a lack of choice, in the high fertility rates of many women.

Countries with greater gender equality also have lower fertility rates than countries with lesser gender equality, although, as we've seen before, those relationships are confounded by country differences in wealth. Poorer people have more children than richer people, in part because of the eventual economic benefits of larger families (more help on the farm, greater assistance in old age). Nonetheless, part of the effect of gender equality is probably the greater decision-making power of women in such countries.

On average, men desire more children than women do, and this sex difference is larger in traditional countries where fertility rates are high. When we consider who bears the brunt of the work in pregnancy and childcare, it's no surprise that women want smaller families than men. As a result, the more control women have over their fertility, the fewer children they have. Again, these data suggest that the decreased fertility that accompanies female education is a clear example of education increasing autonomy.

Education Is an Exercise in Delayed Gratification

In the biblical story of the Tower of Babel, the descendants of Noah decide to build a giant tower in what appears to have been an affront to God himself.* Displeased by their plans, God confused the workers by causing them to speak multiple languages and scattering them across the globe. This biblical tale might have been a disaster for the ancient Babylonians' construction plans, but the outcome it represents was an enormous boon to science, as different languages provide a window into what people of different cultures are thinking. If a language has a word for a concept, we can be pretty sure that members of that culture think about that concept, potentially giving us insight into their psychology. For example, the German word *Schadenfreude*—joy in someone else's sorrow—gives us a peek into German psychology. It tells us something about American psychology, too, given that we resonate so strongly to the word (even though I've never met anyone who knows the English word for the same feeling: epicaricacy†).

Edward Sapir and Benjamin Whorf turned up the volume on this idea that language gives us insight into the speaker's psychology

---

*Some ancient texts suggest that the tower was intended to give people a safe haven should God decide to inundate the Earth again.

† I'd never heard of it either; I googled it.

by proposing that language constrains thought. According to Sapir and Whorf, if I have a word for Schadenfreude I can experience it, but if I don't have such a word, I really can't experience it (at least not fully). This hypothesis was hugely influential in the 1930s but fell out of favor for lack of good evidence. The problem was that cultures that had lots of words for things typically had lots of experience with those things, making it difficult to know if experience led to words or words led to experience. Do the Inuit have more words for snow because they encounter and think about snow all the time, or does their extensive snow vocabulary help them think about it?

Enter Lera Boroditsky. She was fascinated by the hypothesis that language influences thought and had a great idea about how to avoid the chicken-and-egg problem that commonly plagued this line of research. Rather than compare cultures that have fewer or more words for the same thing, she chose to study languages that described the location of events by reference to one's own body or to the Earth. Because all humans have bodies with a front, a back, a left, and a right, and because all humans live on our planet with an east, west, north, and south, the linguistic choice of using one set of terms over the other allows researchers to escape the differential experience problem. Any culture that uses *left* or *right* to say where something happened could have described the same event in terms of *east* or *west* and vice versa. The beauty of Boroditsky's approach lies in the fact that some languages (like Kuuk Thaayorre or Guugu Yimithirr, which are spoken about thirteen hundred miles north of me in my adopted state of Queensland) have no words for left or right or front or back and only identify locations in terms of cardinal directions.

When we test speakers of such languages, we find they're perfectly capable of thinking about things outside their vocabulary. For example, if they're asked, "OK, you say the ibis that stole your lunch was to the north of you. Was that on the side of you with a face or the side of you with a butt?," they can readily answer the question. In other words, Boroditsky shows us that French people also experience Schadenfreude, despite having no word for the concept.

But here's the kicker; Boroditsky also found that people who describe the world in certain ways become expert at coding the world that way. From early childhood Kuuk Thaayorre and Guugu Yimithirr speakers effortlessly know which direction is north and easily remember the cardinal directions of events they witnessed. People who use left and right often struggle to know which direction is north (or if they're like me, they have no clue) and have little hope of remembering that information when reflecting on events later. Boroditsky's research shows us that Sapir and Whorf were on the right track but were looking at the problem from the wrong side. Language does shape thought, but not by limiting what we can think when we lack a word. Rather, language makes us expert in a domain if the way we use language continuously brings that domain to mind.

How do we extrapolate from these data to understand the impact of education on autonomy and connection? Boroditsky's research suggests that a life that de-emphasizes tomorrow doesn't prevent you from thinking about it. And indeed, hunter-gatherers who live in immediate return societies do not save food for tomorrow, but they do engage in extensive preparations for tomorrow as they think about the hunt, make plans with friends, etc. Thus, they are perfectly capable of planning for the future when they need to do so. Nevertheless, Boroditsky's research also shows us how everything changed once we shifted to a world of agriculture and were forced to think about the future on a regular basis. Like people who constantly think about cardinal directions because they describe the world in those terms, we have become experts in tomorrow.

Evidence from a variety of domains is consistent with this interpretation, but my favorite comes from a recent field experiment by Hamidreza Harati and Thomas Talhelm. They wondered if chronic

water shortages might change cultures by requiring people to continually keep long-term concerns in mind, thereby orienting them
away from the here and now and toward the more distant future. To
test this possibility, they compared people in the Iranian desert city
of Yazd, which gets only two inches of water each year, with their
fellow Iranians who live a few hundred miles southwest in Shiraz,
which gets about a foot of water each year.

To test whether people in these two cities were more chronically oriented toward the here and now versus the future, they
posted job ads for a software company that was either described as
stable and long term or an exciting start-up. The stable company
was hiring a full-time position with job security and the start-up
was hiring a part-time position with flexibility. Sure enough, people in humid Shiraz were more likely to apply for the exciting
job with flexibility, while people in arid Yazd were more likely to
apply for the full-time position with job security. Despite the fact
that water shortages have absolutely nothing to do with careers at
a software firm, job security, or work flexibility, the chronic future
orientation of the people of Yazd made them more interested than
the people of Shiraz in finding a job with long-term stability and
security. As Boroditsky's research on language shows us, our environments create *habits of thought* that become ingrained, influencing our approach to life even in domains that are unrelated to the
environmental pressures that created those habits.

Our modern information economies and the education they require have moved us well past the impact of agriculture by making
our entire childhood and adolescence a lesson in delayed gratification. The goal of education is to provide us with the information we
need to thrive in a knowledge economy, but the unintended consequence is that school chronically forces us to prioritize tomorrow
over today. From a very young age, we are taught to study rather
than spend time with our friends so we can pass the test, get into
college, get an internship, or get a good job. Even when children are

too young to have any homework, school demands they spend their time learning math and English rather than running around outside with their friends. Recess and time with friends is something children earn after completing a block of work and is often taken away if they misbehave. In short, school forces us to think about later today or tomorrow on a daily basis, starting in kindergarten and not stopping until we graduate high school, college, or with an advanced degree.

This mandated focus on the future has clear psychological costs, the most notable of which is the price we pay in connection. Due to the centrality of our connection needs, our mind naturally drifts to thoughts about how to create and consolidate connections with people in our lives. School might as well have been designed to beat that habit out of us. By spending our entire childhood learning to sacrifice connection in service of competence (an autonomy goal discussed in Chapter 3), school slowly shapes our pattern of thought until we can't help but think that our foremost goal is autonomy with connection playing second fiddle. School may give us valuable lessons in how to earn a living, but it simultaneously teaches us to make costly sacrifices in our relationships that undo much of the happiness we gain from the income and autonomy education provides.

Before we turn our attention from this topic, you're probably thinking that your school experience was a pretty social one. Mine certainly was, and even though I hated getting up at six o'clock and trudging to school through the snow (and in the dark; remember I grew up in Alaska), I loved catching up with my friends throughout the day. The opportunity to get together with friends and meet new people is one of the major reasons why school is better than YouTube, even though you can probably learn most of the things you need to know on YouTube with less effort and expense. My point is not that school is devoid of social experiences, but that the formal aspects of schooling undermine connection. The informal aspects of schooling and people's motivations for being there often put connection back in the driver's seat, but the formal demands of school are in near-permanent conflict with these social desires.

Done and Undone by Science

In the late nineties I had the good fortune to attend a talk by Ron Suskind and Cedric Jennings. Suskind had just published his wonderful book *A Hope in the Unseen*, in which he documents Jennings's remarkable life. Raised by his mother while his father served a prison sentence for drug dealing, Jennings overcame an impoverished background to excel in high school and eventually attend an Ivy League university. I was keen to hear his story and was not disappointed; I've never met a more positive person in my entire life. After their talks they took questions from the audience, at which point a woman asked Jennings how he could remain so optimistic when he had experienced so much tragedy. Jennings replied that he knew Jesus would never put a barrier in his path that he couldn't overcome. Every time something went wrong in his life, as it often did, his job was to figure out how Jesus intended for him to solve it. He knew the answer was there; he just had to find it.

Jennings's approach to life may be the world's most effective pathway to psychological resilience and success, as every barrier he encounters becomes a challenge rather than a threat. His philosophy also imbues his life with meaning and gives him connection to a larger purpose. Science has taken that away from many of us. Our knowledge that we are just another species on an insignificant planet in a small solar system on the edge of one of many billions of galaxies might seem irrelevant to our connections, but our insignificance calls the very basis of our connections into question. As we discussed in Chapter 5, our ancestors believed they were part of an unbroken chain that stretched for all time in both directions. In one sense, they were right. We can trace our ancestry back to the first life on this planet, and we are all brothers and sisters in that sense. But while science has confirmed this one key aspect of our place in an unbroken chain, it has called many others into question.

Our ancestors believed their ancestors became spirits who could help them through life, and as they prayed to these ancestors for guidance, they understood that someday they too would become spirits who would look out for their descendants. Most of us no longer believe this to be true. Our ancestors also believed they were "the people," chosen to make a living where and how they did. The science of ancient human migrations strongly suggests otherwise. Lastly, our ancestors believed they sat at the center of the universe with the heavens spinning above them. Copernicus showed us that our planet is not the center of anything, and Darwin finished the job by showing how we arrived through mindless processes rather than grand design. As Darwin famously noted, "there is grandeur in this view of life," but it's not the grandeur we thought it was.

Darwin's theory and subsequent research in genetics explain our connection to all other life forms on this planet, and it's fascinating to ponder our relationship with other species, such as our distant cousins the scallops or our even more distant cousins the parsnips. But even though I find it interesting that I share so many genes with my dinner, I'm not sure that the relationships documented in phylogenetic trees help me understand my purpose or meaning in life. I'm not entirely sure I have purpose or meaning in life.

Loss of purpose and meaning is costly to happiness. As you might recall from Chapter 6, religious participation is associated with greater happiness. One potential reason for that relationship is that religion might lead to happiness because it imbues life with connection to a higher purpose and meaning. We can explore that possibility by taking a deeper dive into the religion data. Here I've plotted the happiness levels of people in the General Social Survey as a function of whether they never pray or pray at least some of the time. Because prayer predicts attending religious services, which we know is a source of connection and happiness, in this graph I've only plotted happiness levels among people who never attend services, but either pray or don't.

People who never attend services are of two types: those who don't go to religious services because they don't believe in God and those who don't go to services despite their belief in God (perhaps they don't like religious institutions or they live too far from their nearest church). Thus, any happiness differences that emerge as a function of prayer among people who never attend services are likely to be brought about by belief in God in its own right. As you can see, the effect of prayer is not terribly large, but it's there; people who pray are more likely to be very happy and less likely to be unhappy than people who don't.

Religion is far from the only way to find meaning or purpose, but almost everyone in the group who prays probably feels their life has meaning and purpose (reflected in their relationship to God). In contrast, it's likely that many of the people who never pray feel their life has little or no meaning or purpose. I suspect our ancestors never searched for meaning because they believed they had it. Now many of us feel we don't.

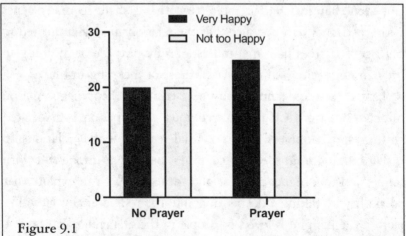

Figure 9.1

Percentage of Americans who never attend religious services but either do or do not pray who are very happy or not too happy.

When I was six years old, my father explained to me how the sun is essentially a hydrogen bomb that is so vast that it explodes continuously. I was fascinated to learn that the sun was an exploding bomb but worried about the implications of what would happen when it ran out of fuel. He suggested that I shouldn't worry about it, as it has a good five billion years of fuel left. That reassured me but led me to ask what would happen then. He said that at that point the sun would explode or collapse and all life on Earth would cease. In that case, I asked, what's the point of all this? He just shrugged.

I doubt I pondered the issue for long, as a game of kickball or Monopoly* likely captured my attention. But although my six-year-old mind didn't give the matter much thought, lack of connection to a higher meaning or purpose can be debilitating. People who feel their lives are without meaning or purpose aren't as happy as people who have meaning and purpose, which is one of the reasons why searching for meaning is associated with unhappiness; the people who are looking are the ones who don't have the answers. Somewhat surprisingly, however, the process of looking for meaning makes most people feel better, which suggests that we may find answers when we search for them. If so, I suspect those answers are often about connection—life has meaning when we take care of one another. But that doesn't change the fact that searching for meaning in life brings our autonomy to an entirely new level that would have been inconceivable to our ancestors.

Freedom is the basis of autonomy, but by removing our understanding of our connection to the universe, science has given many of us too much freedom. With freedom comes the responsibility of getting it right, which can be an extraordinary burden given the scope and magnitude of these questions. For many

* This was 1969, three years before the video game Pong was invented. Monopoly was a go-to game.

people, this level of autonomy produces existential dread rather than exhilaration.

A common solution to this problem is to try to do something meaningful in life. There are numerous ways to attain this goal, with effective altruism (the goal of helping as many other people as efficiently as possible) being a relatively modern approach that has attracted a lot of adherents. I suspect that many people seek meaning via their careers. As we discussed in Chapter 7, one of the primary forms of self-expression since the advent of cities is choosing a career. In that sense, choosing a career is no longer just about figuring out how to make a living; our goal has been transformed into finding and pursuing our passion. That passion then defines us and gives meaning to our existence.

The chance to pursue your passion is an enormous opportunity, but it is also a significant burden. Young people are expected to discover their passion among a seemingly endless array of possibilities, a process that is both daunting and confusing to those who don't have an obvious talent to develop. With so much hanging on this choice, the existence of so many options is intimidating. By way of example, imagine you walked into an upscale chocolate shop. Laid out in front of you are dozens of different chocolate truffles. They can be distinguished from each other if you look at them closely, but how many chocolates can a person reasonably inspect (particularly while the line grows behind you and the shopkeeper becomes increasingly agitated)? And how are you to know which choice is right for you without trying all of them, a process that would assuredly make you sick?

Somewhat ironically, when faced with so much choice, people often do a poorer job picking what they like than when there are only a few options. If asked to choose from chocolate, vanilla, and strawberry, most people can pick what they prefer most of the time. But if they are asked to pick among hundreds of flavors, most people are at a loss and are just as likely to make a poor decision as they are to make a good one. Which of these hazelnut pralines do I like best, and how do they compare to the macadamia options on the shelf above them? I have no idea. The upside of this level of autonomy is the opportunity to choose a career that is a perfect expression of your identity, but the downside is the cost of making the wrong choice when faced with the single biggest decision you've ever made.

The existence of so many career options in a world without meaning can be simply overwhelming. If we still had the strong connections that characterized our lives until the last few generations, we would have plenty of people who could advise us through these complex decisions. The presence of such advisors might seem to limit our autonomy, but seeking or following advice doesn't inherently impinge on your autonomy. Rather, your ability to seek and follow advice *of your own accord* is another expression of autonomy, even though it relies on your connections for successful execution. But as I've argued throughout this book, a world that provides so much autonomy is also a world that prunes our connections, with the result that we lack the connections that allow us to fully engage with the extraordinary autonomy that is now available to us.

Education for the Knowledge Economy Enhances Autonomy

After decades of lower court skirmishes, a 1972 Supreme Court decision (*Wisconsin v. Yoder*) confirmed the right of the Amish to pull their children from school after completion of the eighth grade. The habit among Amish communities of ending formal education before high school was widely regarded as a way of preventing their children from being exposed to ideas that may challenge their faith, such as evolution. But because the Amish are generally employed in crafts and trades, it makes sense for their children to spend less time preparing to join the knowledge economy and more time learning their trade. There is probably a fair bit of truth in both reasons for ending their education at the onset of adolescence, but there's more to it than that. By ending their formal education at such a young age, the Amish minimize the degree to which school damages the strong connections and interdependence that are central to the success of their tight-knit communities.

Raising a barn by yourself without power tools or heavy equipment is nearly impossible, but in Amish communities everyone pitches in and they get the job done in a day or two. The same holds for bringing in the crops, or almost any other large-scale and difficult farming activity in which the Amish engage. Tasks such as these can be accomplished through the close cooperation of many skilled participants with very simple tools, or they can be accomplished on your own with some fancy power tools and heavy equipment. Even the complete novice Jeremy Clarkson can plow a large field in a single day if he sits behind the wheel of his massive Lamborghini tractor.*

Like the technology that enables the clueless Clarkson to farm on his own, education creates islands of self-sufficiency. Manual labor requires that people work together in the same physical space. If you and I are building some sort of structure or machinery, it helps to have us both in the same place at the same time, which is why construction sites swirl with activity. In contrast, mental labor can almost always be done at a distance and asynchronously. If you and I are writing a computer program or a movie script, we might even be more productive if we're in different time zones so you can work on my ideas once I hit the hay.

The COVID pandemic was particularly hard on blue-collar and service workers because they need to be on-site simultaneously to conduct their business. Workers in the knowledge economy, in contrast, emerged from lockdown relatively unscathed. They (and

---

* See *Clarkson's Farm* on Amazon Prime.

their employers) learned that they could do most of their work from home, that they didn't need to be co-located with their colleagues, and that it often didn't matter if they conducted their work during typical business hours. Participation in the knowledge economy involves far less connection than participation in service or manufacturing.

The island of self-sufficiency you become when your education prepares you to work in the knowledge economy is a costly achievement. On the positive side, if you're a knowledge worker, you probably didn't lose your job when the pandemic hit; I'm not sure you even noticed there was a pandemic. On the negative side, a job spent alone at home staring at a computer screen is not what we evolved to do. That's snow leopard work. By training you for such occupations, education not only enhances your autonomy, it undermines your connection. For better and for worse, the knowledge economy and the many years of education required to join it transform us into anti-Amish.

Wealth Enhances Autonomy

Poor people need each other, rich people don't. When poor people require a tool they don't own, buying it is often out of the question but borrowing it is not. When poor people need someone to watch their dog, hiring a dog-sitter is out of the question but asking a neighbor is not. For these and a thousand other reasons, poor people depend on each other. People in poor neighborhoods live in a complex web of interdependence, with everyone counting on each other to help keep the boat afloat. In this sense, the connections among poor people are not that different from those of our ancestors; poor people form tight reciprocal relationships out of necessity.

As we've discussed, tight connections have costs. Because favors must be repaid, if you don't need help, you might think twice before asking for it. Rich people are far more likely to buy the extra tool, hire the dog-sitter, or pay for the help they need than they are to rely on each other. Depending on how much you like your neighbor, you might prefer your latte to be delivered by drone rather than pop over to your neighbor's house when you run out of your favorite beans. It might be perfectly pleasant to chat with your neighbor now, when you want that coffee, but who knows when your neighbor will meander back to your place in search of a reciprocal cup. Because connections undermine autonomy, constraining your behavior in often unpredictable ways, people are far more likely to form them out of necessity.

We see these effects of wealth in several different ways. First, we can go back to the General Social Survey and ask how often Americans get together with their neighbors as a function of income. When we plot these data in Figure 9.2, we see that poor people are about twice as likely as rich people to get together with their neighbors at least a few times a week. In contrast, rich people are about twice as likely to see their neighbors only a few times a year. Socializing with your neighbors might seem like a cost-free activity, in the sense that you're not borrowing tools or asking them to dog-sit. But it takes time to get together with people, meaning that you're more likely to do so when you benefit in other ways by maintaining strong relationships.

The same effects emerge as a function of education. As you can see in Figure 9.3, people with only a tenth-grade education are the most likely to see their neighbors several times a week and the least likely to see them several times a year. People with a graduate degree are the exact opposite. Education is correlated with wealth, so it's not easy to disentangle which one plays a more important role, but these data show us that as people become educated and wealthy they pull away from their neighbors.

It's important to note that these patterns don't emerge when we examine the time people spend with their friends. Once we control for age (educated and wealthy people tend to be a little

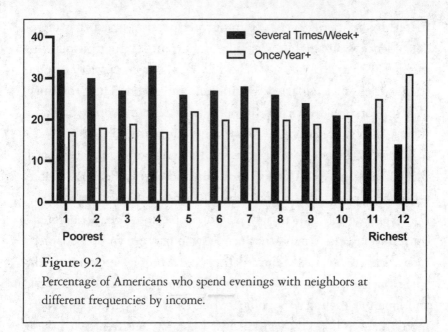

Figure 9.2
Percentage of Americans who spend evenings with neighbors at
different frequencies by income.

older than average), we see no effect of education or wealth on time
spent with friends; only time with neighbors suffers. Wealthy and
educated people pulling away from their neighbors and not their

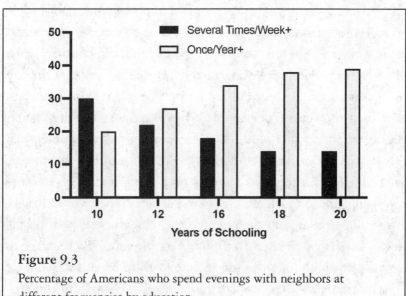

Figure 9.3
Percentage of Americans who spend evenings with neighbors at
different frequencies by education.

friends makes perfect sense from our perspective, given that interdependence with neighbors is more important for poor people who might not have an easy way to borrow things from possibly distant friends, babysit their friends' children, and so on. The proximity of neighbors makes it more important that poor people maintain their relationships with them for pragmatic reasons.

These findings are consistent with a series of studies by social psychologists Michael Kraus, Paul Piff, and their colleagues, whose research paints a similar story when it comes to money and connection. For example, when they placed people into "get acquainted" conversations, they found that people who had grown up poor were more attentive to each other than people who had grown up rich. Their poor participants were more likely to make eye contact, nod, and laugh at each other's points, while their rich participants were more likely to doodle or check their phones. Consistent with these behavioral differences, poor people were also more likely to take each other's perspective, which helps them read each other's emotions.

Their experiments also revealed that when people were reminded of all the ways their future was unpredictable, poor people became more communally oriented than rich people, apparently because poor people are more likely to need the help of others when life gets dicey. Poor people chart a route through chaos by building and maintaining connections, while rich people rely on money to solve their problems. For example, when asked whether they would take a promotion that required them to move to another city, rich people were more likely than poor people to jump at the chance, particularly when they were considering a chaotic future. Because poor people have much less disposable income, they would benefit more from a promotion than rich people. Nonetheless, poor people view connections to their community as the most likely factor to see them through tough times, and are loath to break such connections, particularly when considering a possibly chaotic future.

These findings from the laboratory help explain the differences we saw in the last chapter, where people in agrarian countries valued connection more and people in industrialized countries valued autonomy more. The average yearly income across the industrialized countries was a little over fifty thousand US dollars per person, while the average income in the agrarian countries was slightly less than six thousand US dollars per person. These results provide further evidence that wealth enhances autonomy and decreases connection. Because rich people no longer have pragmatic reasons to connect, they allow their autonomy needs to take precedence over their connection needs.

It's worth keeping in mind that causality can also go the other way. Part of the reason poor people are focused on connection is because they need to be, but perhaps people who are more focused on connection end up poorer because they refuse to make the sacrifices in connection that would allow them to become rich. For example, many people choose careers in areas like primary and secondary school teaching because they value the connection that is inherent to the job, even though teaching is relatively low paying for the amount of education it requires. All else being equal, people who value connection are likely to end up poorer than people who value autonomy. As is often the case, I suspect causality goes both ways.

In combination, these data provide compelling evidence that rich people are less focused on connections because they believe they can get by without them. Possibly they're right. Perhaps wealthy people pay no psychological costs for noodling on their phone rather than chatting with the person they just met, or for skipping the neighbor's barbecue. People presumably know how they want to spend their time and perhaps these data simply reflect different life choices as rich people pursue other opportunities. That's certainly possible and I suspect it's often true, as some people are happiest when they maintain only a small circle of social relationships. But I also suspect that it's often untrue; most people are happier when circumstances push them to socialize.

We can test this latter claim by diving back into the religion data from Chapter 6. As we discussed, religious participation makes people happier, particularly when it involves attending religious services. This latter finding suggests that the social aspects of religion play an outsized role in life satisfaction. Cross-cultural research also suggest that religion protects people from some of the psychological costs of poverty, as religiosity has a larger positive effect on life satisfaction in poorer rather than richer countries. This finding makes intuitive sense, as it's easy to imagine that belief in a higher power is more comforting for people who are struggling to get by than for people who have it easy. If you recall Cedric Jennings, his belief in a benevolent God played a pivotal role in his ability to succeed in remarkably difficult circumstances.

But it's possible that there's more to the story than that. Rich people might not need religion to believe that everything is going to be okay, but they might need religion to nudge them into connecting when they have few pragmatic reasons to do so. According to this possibility, belief in a higher power might do more for poor people than rich people, but attending religious services might do more for rich people than poor people. To test these possibilities, we need to separate the belief aspects of religious experience from the social aspects. Although we can't do that perfectly, we can use the General Social Survey to get pretty close.

To assess whether belief in God, *on its own*, is differentially important for the happiness of rich and poor people, we can take another look at the effect of prayer on the happiness of people who never attend services. We saw earlier in this chapter that prayer among such people is associated with happiness, but this time we can separate the data by whether people are poor or rich. When we do so, we see that believing in God (as indicated by whether people ever pray) is slightly more important for the happiness of poor Americans than rich Americans (Figure 9.4). Setting aside the unsurprising fact that rich people are happier across the board, we see that the proportion of very happy folks rises more and the proportion of sad (that is, "not too happy") folks drops more among the poor than the rich when people pray. Because we're only considering people who never attend services, these larger effects of prayer among the poor than the rich suggest that belief in God on its own is more beneficial for poor people than rich people.

We can't isolate the social aspects of religion quite as cleanly in this dataset, as attending religious services is also a marker of believing in God. But we can ask whether attending religious services has a bigger effect on life happiness among the rich than the poor. As you see in Figure 9.5, the answer is complicated. On the one hand, going to services increases happiness more among the rich than the poor. On the other hand, going to services decreases sadness more among the poor than the rich. These data suggest that

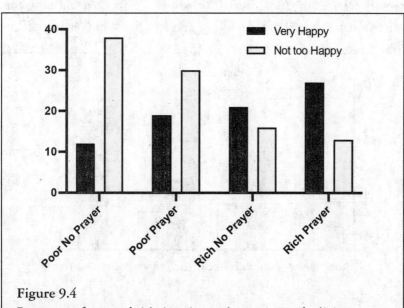

Figure 9.4
Percentage of poor and rich Americans who never attend religious services but either do or do not pray who are very happy or not too happy.

going to services protects poor people from sadness and provides rich people with the extra boost they need to be very happy.

Just as we saw earlier, the impact of attending religious services is substantial. When rich people attend services several times a week, they're about twice as likely to be very happy as their rich peers who never attend services. These findings suggest that attendance at religious services leads rich people to form and maintain social connections they are otherwise inclined to avoid. In so doing, they also show us the power of connection, even when we're only connecting because our circumstances force it upon us.

When we compare the results of prayer versus attendance at religious services, we see that attending services has a much bigger effect on happiness than prayer does. Indeed, the effect of attending services is so large that *poor people who go to services several times a week are happier than rich people who never go at all.* That's such a remarkable result that I initially thought I'd made a coding error, but it's true—regularly attending services has a bigger impact on

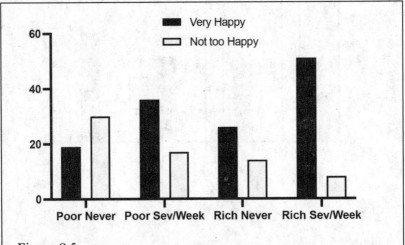

Figure 9.5

Percentage of poor and rich Americans who are very happy or not too happy by whether they never attend religious services or attend several times per week.

your happiness than wealth. Money buys a fair bit of happiness but connection gives you more bang for the buck.

City living, education, and wealth are all evolutionarily novel, in the sense that none of them existed during most of the time humans have lived on this planet. Cities and formal education are very new—only a few thousand years old—and wealth is just a little bit older, with agriculture being the primary driver of wealth in most societies. Cities, education, wealth, and science have all improved the human condition to such a degree that it's hard to imagine life without them, and many of their benefits have manifested in increased autonomy. Unfortunately, you can't increase autonomy without paying a price in connection, and connection is critical for life satisfaction.

I suspect that wealthy, educated urbanites are paying a steeper price for their lifestyle than they realize. Indeed, I believe that many of us have paid too great a price in connection for our increased autonomy and it's time to shift back toward the tighter connections of our ancestors. Before considering this possibility, however, we need to turn our attention to marriage to see how it has served as a bulwark against the erosion of connection while also exacerbating the problem.

10

Marriage and the Struggle Between Connection and Autonomy

Marriage is one of the strongest forms of human connection, so it's no surprise that the conflict between autonomy and connection is amplified and reflected in people's marriages. The earliest evidence of an official marriage is over four thousand years old from Mesopotamia, but de facto marriages are probably as old as humanity itself. Marriage, or the formation of long-term romantic bonds akin to marriage (from here on I'll simply refer to such relationships as *marriage*), is universal in hunter-gatherer societies. We don't know how prevalent partner choice was in the pre-history of marriage, but anthropologists who live(d) among people of different cultures report that in most hunter-gatherer societies, the parents (and sometimes other relatives) play a key role in who marries whom. The two people who are matched up can exercise their opinion, but their preferences are often secondary to those of their parents or older relatives.

Divorce was probably invented a few months after the first marriage. Divorce, or more frequently the unceremonious ending of the relationship, is common among hunter-gatherers. People in such societies tend to be serially monogamous, although some relationships last. For example, among the Hadza—who choose their own partners without much family interference—their ethnographer Frank Marlowe estimated that about 20 percent of marriages last for life. Once marriage transitioned to a formal event in agricultural societies, however, the easy form of divorce that the other 80 percent of the Hadza relied on all but disappeared.

Hunter-gatherers live in extensive cooperative networks that are entirely fluid. Everyone in camp cooperates with everyone else and anyone can leave to join another camp at any time. Most agricultural societies differ from hunter-gatherers in both of these ways. People only share with close friends and family and no one leaves because they've invested enormous effort in their land, often for generations. Due to these differences in how society is structured, marriage has ripple effects among agriculturalists that are absent among hunter-gatherers. In hunter-gatherer groups, marriage has little impact on anyone other than the married couple (and those who might have been hoping to partner up with one of them). In agricultural societies, marriage changes the structure of both families by extending the group that is now kin. By creating new cooperative and financial relationships, with implications for extended family on both sides, marriage in agricultural societies alters the obligations and opportunities for both families. As a result, ending marriages was not a simple matter for agriculturalists.

That's not to say that affairs weren't common in agricultural societies, just as they are in all ostensibly monogamous species on this planet. Rather, once societies codified marriage into law, they didn't look kindly on dissolving the bonds that linked so many people together. People were officially married for life regardless of their actual behavior on the ground. The purpose of marriage was no longer just the recognition of a close relationship between two people that facilitated child-rearing, as it was in many hunter-gatherer communities. Rather, marriage became an economic bond between families.

As we discussed in Chapter 8, prior to two hundred years ago, over 90 percent of humanity lived in rural areas and engaged in some form of primary production like farming. In such an environment, marriage extended the cooperative networks among people who were at constant risk of falling into abject poverty if crops failed or some other disaster occurred. When you reflect on the demands of such societies, the seemingly draconian but nearly universal laws that prevented divorce become much more sensible than they might otherwise appear. With the move to cities, however, the bonds between families diminished in importance as people became less dependent on each other, and marriage started to take on new meaning. In the United States, the migration to cities began comparatively early, with urbanization accelerating in 1840 in the US and not until 1900 in the rest of the world.

By 1850, marriage in the US was meant not just to be based in affection, but ideally to be preceded by it. Love became a precursor to marriage rather than something people hoped would grow once they were together. Marriage was still a utilitarian enterprise, but utilitarian concerns were no longer in the driver's seat.* It would take over a hundred years before divorce law caught up to the idea that a marriage based in love should end when love dies, with the advent of no-fault divorce in California in 1969 and elsewhere in the US in the seventies and eighties. Nevertheless, the meaning and purpose of marriage changed as Americans moved from the country to the city.

The next big redefinition of marriage took place as Americans became rich enough to start withdrawing from their community, but in this case the change in marriage was both cause and consequence of this loss of connection. As Americans started separating themselves from friends and neighbors, they achieved greater freedom

---

* At the risk of straining the metaphor, I'd say that Romance was driving and Utility was riding shotgun.

but lost reliable and meaningful social support. Spouses were well suited to fill that void, in part because they're meant to be soulmates and in part because once you're married, you might as well lean on your partner for additional comfort and support. As a consequence, the decline in time spent with friends over the last fifty years is even more pronounced among married people than among single people, although the effect is evident among both. Thus, on the "marriage leads to disengagement from others" side of the equation, married people have been even more likely than single people to opt out of their network of connections over the last few generations.

In principle this corrosive effect of marriage on relationships with friends and neighbors can be offset by greater spousal closeness. And for many it is. But as Eli Finkel and his colleagues point out, if you hope to gain additional benefits from your marriage you need to put in additional effort, and Americans are not putting in effort commensurate with the rewards they seek. Indeed, Americans are spending less quality time with their spouses now than they were fifty years ago, if by quality time we mean time spent alone together. This effect is more noticeable among couples without children, largely because parents with young children have never had any alone time in the history of our species. When we drill down into the details, we see that spouses are less likely to do things as a couple than they were fifty years ago, such as eating, engaging in leisure activities, or even visiting friends together.

Work appears to be part of the problem, as spouses report spending more time working than in the past, but there's more to it than that. Perhaps it's simply the case that the enormous number of entertainment and exercise options that now exist make spouses more likely to go about their separate lives. Whatever the cause, the combination of less time together and higher expectations placed on marriage is associated with declines in marital satisfaction over the last fifty years. Thus, on the "disengagement from others leads to changes in marriage" side of the equation, married people expect more from their marriages now that they have severed many of their ties with friends and neighbors, but those expectations often remain unfulfilled for lack of a commensurate increase in marital investment. As a result, we see declining connection both inside and outside the home, as people appear more inclined to do their own thing (a strong form of autonomy).

The Rise in Single Living

My friend Peter is six-foot-four and full of muscle.* When he was in grad school he used to show up to lab meetings with his lacrosse gear over his shoulder and a few tufts of grass in his hair, having just come from practice. If you look him up on the web, you'll see that he's a business school professor, an author, and a fierce advocate of single living.† Given that he's the kind of guy almost every woman swipes right for, that might not seem surprising. Why should he settle down when he's inundated with offers? But it's more complicated than that. Peter is a spokesman for the growing solo movement— people who forgo marriage and other forms of lifelong commitment to just one person. In the United States, we have now reached a point where approximately half of single adults are not very interested in casual dating, much less finding a long-term partner.

The movement away from marriage has been growing for quite some time. Marriage rates have dropped steadily in the United States for the last fifty years, just as they have in much of Europe as well. What explains this decline in marriage and what does it tell us about the changing balance between autonomy and connection? As you might imagine, when societal effects are large and widespread, there tend to be lots of causes. First, having children outside

---

* Six-five actually—I just like that line from the Men at Work song, "Living in a Land Down Under."

† His latest book is *Solo: Building a Remarkable Life of Your Own.*

of marriage has become normative in many countries, particularly in the US and Northern Europe. These data suggest that one reason to get married—to have children—is no longer important in most places (but not all—having children outside of marriage remains very rare in East Asia). Increased wealth and technology have made having children outside of marriage a manageable proposition for many people.

Second, it's possible that many people are forgoing the formality of marriage but nonetheless forming long-term relationships with each other, hunter-gatherer style, and the data are consistent with that possibility as well. But that's not the whole story. Rates of marriage-plus-cohabitation have dropped slightly worldwide, and dramatically in North America and Northern Europe. More Americans (and Europeans) are living alone than ever before. In 1850, one in every hundred American adults lived alone. By 1950 that number was one in twenty-five and today it's one in seven. That's an enormous change in the number of people going solo. Necessity connected every single one of us when we were hunter-gatherers. As recently as 170 years ago, it still connected 99 percent of us. Those days are gone.*

These data implicate major societal changes in the US and Northern Europe that are causing long-term relationships to fall out of favor. Although it's difficult to know with certainty what those changes are, most analysts attribute the rise in solo living to women's increased participation in the labor force as well as increased gender equity more broadly. Men are still more likely than women to live alone (excluding old age, as women typically outlive men), but economic independence has freed women—and hence men—from the need for marriage and other forms of long-term relationships. Women no longer need men to support them, which

---

* As a sign of the changing times, *The Economist* recently created the "Carrie Bradshaw Index," which tracks the price of living solo in the US (New York was the most expensive city and Wichita was the cheapest).

along with better birth control has given both parties greater sexual and social freedom. Economic power is the key to this increased self-reliance, and America and Western Europe have become over three times richer in real terms across the last sixty years.

The fact that long-term relationships have fallen in lockstep with rising wealth and gender equality suggests that *people stop marrying when they don't have to*. When survival and creature comforts no longer demand connection, people begin to weigh autonomy more heavily in their life choices. We saw this first in the shift from hunter-gathering to agriculture, where connections between unrelated group members began to diminish as group-level sharing was no longer required for survival and hence no longer mandated. We saw it again with the move to cities and increased education and wealth, where connections with neighbors and even friends were no longer necessary and began to diminish. And now we're seeing it one more time in the drop in long-term romantic bonds. In all these cases, when people can survive and thrive without forming such bonds, more and more of them choose not to.

Keep in mind that nothing has happened to make marriage less rewarding than in the past. Even when our ancestors were given free rein in who they could marry—which they often weren't— they typically had only a handful of eligible people to choose from. Thus, it seems likely that long-term relationships today are more rewarding than ever, now that we make our own choices and have so many possibilities in who our friends and lovers will be. The fact that people are increasingly opting out of long-term romantic bonds shows us that the conflict between connection and autonomy has shifted relentlessly in favor of autonomy, as connection has become less vital to our survival. But just because we don't need connection to survive doesn't mean we don't need it to thrive.

There are numerous lines of evidence supporting this conclusion, but for a unique perspective on the problem, let's take a look at a recent study conducted by Jade Butterworth when she was my student at the University of Queensland. Jade was interested in the possibility that unmarried men (and presumably many married men as well) aren't having their connection needs met and hence try to satisfy them elsewhere. To test this possibility, she examined the text messages exchanged between six high-end sex workers ($600+ Australian dollars per hour) and over five hundred of their married and unmarried clients.

Jade found that clients could be grouped into two different clusters. One cluster was made up of men who were interested in forming a connection with their sex worker, as evidenced by banter in their text messages, gifts, signs of caring, and repeat bookings. These men were also more likely to book the "Girlfriend Experience" than the "Porn Star Experience," with the former intended to resemble an idealized dating scenario and the latter . . . well, the latter is pretty obvious. Jade also found that unmarried men were more likely than married men to be in the connection cluster. These data suggest that men who aren't having their relationship needs met are inclined to pay for connection experiences rather than just seeking sexual variety.

The porn star experience has been and probably always will be a moneymaker. But the fact that many men book the girlfriend experience tells us something important about human nature. Sure, we evolved to enjoy sex, as did every other animal. But we also evolved to connect; so much so that men happily pay $600+/hour for a sex worker to act like a girlfriend and pretend to connect with them.

Assortative Mating and Female Autonomy

My birth cohort (people born in the early sixties) was the first group of students in the United States in which women attended college at equal rates to men and they've never looked back. Within just a few years, women were attending and completing college at higher rates than men and they still are; 58 percent of college graduates in the most recent classes are female. With three women on campus for every two men, women find it more difficult to form long-term romantic relationships with men in college than they ever have before. But the relationship challenges go well beyond campus.

As we discussed in Chapter 9, education is one of the great sources of autonomy and opportunity and from that perspective, increased female education is an unmitigated plus. Unfortunately, however, this long overdue expansion of female autonomy has cut into women's connections in ways that go beyond just leaving their hometown friends behind. As you can see in Figure 10.1, the General Social Survey shows that as women become more educated, they are more likely to remain single, while men show the opposite effect. These data raise the possibility that female empowerment has been accompanied by a unique loss of connection, as educated women appear to be paying a price in marriage opportunities while educated men receive marriage dividends. But we'll need to take a deeper dive into the data before deciding if that interpretation is true.

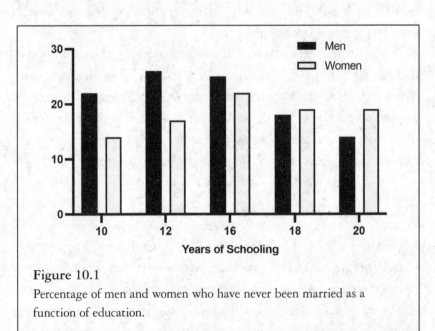

Figure 10.1
Percentage of men and women who have never been married as a function of education.

Figure 10.1 suggests that as women's college graduation rates surpassed those of men, large numbers of women have not found a partner. Of course, this may or may not be a problem. Maybe better-educated women are less willing to settle because they can afford not to, and hence are single by choice. Recent polling data from Pew are consistent with this possibility. When women were asked whether different attributes in a potential partner would make them less likely to date the person, college-educated women were most likely to state that various attributes were a dating liability. Highly educated women are choosier.

Note, however, that a potential partner not having a college education wasn't at the top of educated women's list of liabilities. If we take their answers at face value, educated women are more concerned about whether their potential partner smokes, has a job, supports Trump,* is an anti-vaxxer, lives with his parents, is short (hey!), or is highly religious than whether their potential partner went to college. We don't have good data on all these factors, but we do have good data on education, which suggests that women are slightly understating its importance. Fifty-four percent of college-educated women say they prefer a man who has a college education, but 64 percent of college-educated married women are in fact partnered with a college-educated man. We see a similar effect among men; 44 percent of college-educated men say they prefer a woman who has a college education but 69 percent of college-educated married men are partnered with a college-educated woman. College-educated men and women might be underreporting the importance of education in their relationship preferences, but it's possible they simply spend more time with each other and end up with each other by default.

Whatever the cause, the preference that educated men and

---

* Remember from Chapter 7 that women are more likely than men to be on the political Left.

women show for each other has numerous consequences. Most central to our current discussion, because women are now more likely to go to college than men, they are also less likely to find a partner. Such women are choosing to remain single, but at the same time, their choice is heavily constrained by the available opportunities.*

Marriage Is Changing, but Good Marriages Matter

Not to cast aspersions on my friend Peter's fierce advocacy of single living,† but the connection provided by a good marriage remains one of the most important sources of human happiness. Good marriages are one of the only things scientists have ever found that permanently raise our baseline happiness. There are lots of data that make this point, but my favorite studies come out of Germany where researchers followed thousands of people for well over a decade. With such large samples followed over such long periods of time, it's a virtual guarantee that they'll have the opportunity to track people as they meet that special someone, get married, and eventually either stay married or get divorced. Richard Lucas and his colleagues have taken a deep dive into these data and have made some rather remarkable discoveries.

First and foremost, causality goes both ways. Good marriages make people happy (I'll go through the data in a moment), but

---

* Somewhat ironically, the increased egalitarianism that enabled women to go to college has also resulted in increased societal inequality. Because educated men and women tend to partner up with each other, the marriage market has played an important role in the expanding gulf between rich and poor that we've seen since the 1960s. When educated people marry each other, their choices consolidate resources in highly educated, dual-career families that didn't used to exist. Greater equality of the sexes has come at a cost of greater societal inequality.

† Remember, he's six-five and full of muscle and I'm five-six in thick socks and weigh less than ten stone.

happy people are also more likely to form good marriages. We can see the effect of happiness on marriage by examining people's happiness five years prior to their wedding—at which point they probably haven't even met their spouse yet—as a function of whether they're eventually going to stay together or get divorced. When Lucas and colleagues made this comparison, they found that people who are going to stay together were substantially happier five years prior to their wedding than people who are going to get divorced. These data suggest that happier people are more likely to form good marriages.

Lucas's data also show the other causal pathway, as they found that marriages that end in divorce are associated with notable decreases in happiness every year. Interestingly, the decline in happiness starts the year prior to their wedding, at which point people who are going to get divorced are essentially the happiest they've ever been and ever will be. By their wedding year they're back at baseline, and only one year into their marriage they're already starting to suffer.

What about the other side of the coin; what do these data tell us about the effects of happy marriages? Although I said that good marriages can make you permanently happier, these data don't really make that point. When we look at the marriages in which people stay together, we see a few things. First, the year of their wedding, not the year before, is their happiest year in life. Unfortunately, however, happiness starts to decay in good marriages in the first year as well, although the rate of decay is not as steep as it is in bad marriages. Indeed, four years into a good marriage, people are still slightly happier than they were five years before their wedding. Nevertheless, these data aren't very impressive evidence that marriage makes people permanently happier.

To make that point, we need to dive deeper still and Lucas and his colleagues did exactly that. As you know, people have a variety of reasons for staying with their spouses, only one of which is whether they are in a happy relationship. People also stay with their spouses for the sake of their children, for economic reasons, because they believe in the sanctity of marriage, etc. To get around this problem and look specifically at the effects of good marriages, Lucas and his colleagues divided those who stay together into three groups: people who show the greatest increase or decrease in happiness and folks in the middle.

If we look at the people in the middle, we see essentially the same effect that we saw before, in which people get a little happier when they get married and then slowly drift back to baseline. But the important groups are those at the top and bottom. Here we see that by the tenth year of marriage, the folks who stay together for reasons other than the quality of their marriage (i.e., those on the bottom) are substantially less happy than they were prior to meeting their spouse. They may have decided to stay together, but their unhappiness arc looks a lot like that of people whose marriage ends in divorce.

In contrast, people who are in the top group keep getting happier every year for six years into their marriage, at which point they plateau. Ten years into their marriage, they remain substantially happier than they were the year of their wedding, and if you recall, they were happier the year of their wedding than they were five years earlier. This is striking evidence that a sizable portion of the population can become permanently happier. Good marriages are one of our most important forms of long-term connection, with the result that they have a uniquely powerful effect on life satisfaction. Given the critical role that pair-bonding plays in child-rearing, these data remind us just how central marriage is to the human success story. Evolution rewards those of us who have the good sense and good fortune to get it right in our marriages by giving us a permanent increase in happiness.

Marriage also confers health benefits. Married people live longer than single people, and this effect is more pronounced among men than women. In the United States, for example, a married man aged 65 can expect to live another 18.6 years, but a single man can only expect to live another 16.4 years. A married woman aged 65 can expect to live another 21.1 years, but a single woman can only expect to live another 19.6 years. Women outlive men, but as you can see from these data, marriage is more beneficial for men's longevity than it is for women's. Finally, given the strong links between happiness and good marriages and between happiness and good health, it's no surprise that the health benefits of marriage are stronger among those who are happily married. From gum disease to heart attacks, happily married couples have better health outcomes than unhappily married couples, in part because happy marriages also enhance immune functioning.

These findings tell us a lot about the effects of good and bad marriages, but there's one more thing we need to know: How do married people compare to unmarried people? Based on the data we've seen so far, it would be fair to assume that married and unmarried people have similar levels of happiness. Remember that among the marriages that stay together, some people become permanently happier, some return to baseline, and some become less happy than they were before they married. Thus, when we collapse across all marriages, married people's happiness tends to be pretty much where it was before they met.

But that doesn't mean that married and unmarried people are equally happy. As we saw when we compared people who eventually got divorced to people who stayed together, there were baseline differences in happiness prior to meeting their spouses. Perhaps similar differences exist among people who do and don't marry. So how do unmarried people compare to everyone else? To answer that question, we can look back at the General Social Survey and the happiness levels of people who are married, unmarried, divorced, or separated. As you can see in Figure 10.2, *unmarried people look a lot like divorced people.* Unmarried people are nowhere near as happy as married people; they're about half as likely to be very happy and about twice as likely to be not too happy.* For what it's worth, you can also see that separated people are a little less happy than divorced people, presumably because they are still in the throes of their breakup while for the divorced people their breakup is now in the rearview mirror.

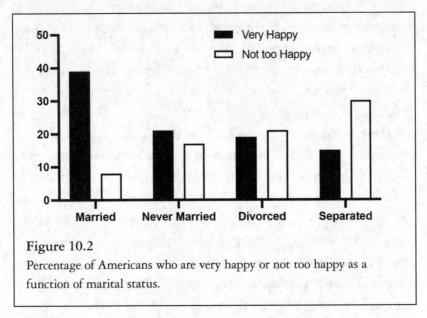

Figure 10.2
Percentage of Americans who are very happy or not too happy as a function of marital status.

What are we to make of these findings? I believe they reveal something important about the balance between autonomy and connection. Because people who get married show an increase in happiness but then return to baseline, these data suggest that people who never marry have lower baseline happiness than people who marry and stay together. We can't be certain why that is, but we are left to conclude that whatever is causing unmarried people

---

* In case you're wondering, these findings are not a function of age; the results look the same when we restrict the sample to different age groups.

not to marry is also causing them to be less happy. As I've argued throughout this chapter, the data suggest that increased autonomy is a key factor leading people to decide against marriage. According to this possibility, *increased autonomy is a major cause of unhappiness.*

I believe that the unhappiness of unmarried people provides further evidence regarding the cost of too much autonomy. There's nothing wrong with high levels of autonomy in principle, but because increased autonomy can only come at a cost to connection, these data show us the price we pay when we sacrifice too much connection in service of autonomy. I believe that unmarried people have overemphasized autonomy and underemphasized connection in many domains of their lives, with the result that their happiness has suffered.

Keep in mind that it's not (necessarily) being unmarried that is making these people unhappy. Rather, their emphasis on autonomy has led to a series of life decisions that have de-emphasized connection, resulting in their unhappiness *and* their being unmarried. We can think of being single as a symptom of a broader problem and not the problem itself. For unmarried people to attain the happiness levels of married people, I believe they need to restore the balance between connection and autonomy across the various domains of their lives. Once that balance is restored, it's possible they would choose to marry but it's possible they wouldn't. As the data with married people suggest, that decision itself will not be critical for their happiness (unless they're one of the lucky folks who finds their perfect match). Of course, this argument doesn't apply to all single people, many of whom are undoubtedly highly connected and very happy. But it does apply to single people on average.

It's difficult to tease apart what's causing what when we study marriage, as the decision to marry involves a huge number of factors, with perhaps the most important one being the mutual nature of the attraction. Because mutuality is so important, and because people don't have complete control over who they are attracted to, every finding has multiple possible interpretations. But if we step back to look at the big picture, a few key facts emerge that tell a complex but coherent story:

1. Marriage doesn't make people happier. This noneffect of marriage is composed of several countercurrents that largely cancel each other out . . .

   A sizable percentage of married people get happier and stay that way.

   A sizable percentage of married people get unhappier and stay that way.

   Most married people who stay together have a spike in happiness when they marry but then return to baseline within five to ten years.

2. Married people are happier than unmarried people.

   It's tempting to conclude that single people are less happy than married people because they can't find the right partner. Before accepting this conclusion, however, consider that an equal number of married people might be unhappy because they defaulted into married life without giving enough thought to whether they'd found the right partner.

3. How could #1 and #2 both be correct? The only way to reconcile these findings is to conclude there must be something about people who choose not to marry that makes them unhappy. I think this *something* is too much autonomy and not enough connection.

4. If we accept this possibility, then unmarried people don't need to get married to be happier, but they do need to prioritize connection in their lives.

We turn now to the final section of the book, in which we consider how to do just that.

Part IV

Rebalancing

Part IV dives into our final question, which is how to rebalance autonomy and connection to bring them closer to their evolved equilibrium and help us regain our well-being. Chapter 11 examines the price we paid when we move our social life online but then considers how we can leverage e-connection in our efforts to rebalance. Chapter 12 discusses how the mismatch between our evolved psychology and our modern environment has led to our current underconnected overautonomous state and considers strategies for reconnecting. Because change is hard—our lifestyle and habits have their own inertia—and because the proper balance will fluctuate across your lifetime, rebalancing is more of an approach to life than a discrete program that will put you back on track. Nonetheless, if you try the strategies laid out in Part IV whenever you find yourself dissatisfied with your state of being, you'll have a better chance of fixing things before they get out of hand.

11

Reconnecting in a Modern World

In his influential book *Bowling Alone*, Robert Putnam laments the decline in service clubs since the 1950s. Many people will no longer be familiar with the Elks and Lions clubs, Rotary and Kiwanis clubs, Freemasons, Shriners, and Optimists, but they were once a major part of American life. Putnam's concern was with the decline in civic engagement that these clubs had previously facilitated, which he argued is necessary for a strong democracy. I suspect he's right, but my concern is with the loss of connection evident in the decline of these clubs. Americans were once connected to each other in myriad ways, but as we've seen over the course of this book, almost all of them are under threat. Most Americans are no longer members of clubs like the Elks or Lions, they no longer spend much time with their neighbors or friends, and they even spend less quality time with their spouses. The inevitable consequence is that they are spending increasing amounts of time alone.

In Chapter 10 we saw that Americans are more likely to live alone than ever before, and surveys show they are also more likely to spend their leisure time alone. Even pre-COVID, Americans were spending almost half of their free time alone. This increase in alone time is not just a by-product of living alone, as it has come at the expense of time with others both inside and outside of people's households. COVID only made the trend worse, with a five-hour per week jump in alone time from 2019 to 2020. It remains to be seen if that increase in alone time will dissipate in our post-COVID world, as many people show a decreased willingness to spend time with groups of friends or in public settings in the aftermath of the pandemic.

These data do not paint a pretty picture. As we discussed at the outset of this book, humans did not evolve to spend their time alone, snow leopard style. We are a gregarious species who depend on each other to maintain our mental and physical health. What we can't tell from these data, however, is how much of that alone time is spent truly alone versus in e-company.

When Connection Went Online

MySpace was created in 2003, Facebook in 2004, Twitter in 2006, the iPhone in 2007, and Zoom in 2011. These platforms, and many others, represent key steps in the progression of e-connection as we moved our social lives online. But e-connection is a lot older than social media. The telegraph was invented in the 1830s and the first transatlantic cable was laid in the 1850s, e-connecting America to Europe for the first time in history. Although most people didn't use the telegraph just to chat, telegraph operators did. When they weren't busy sending paid messages, they gossiped with each other and told each other stories, even though most of them had never met face-to-face. Alexander Graham Bell's modification of the telegraph into a telephone in the 1880s turned this difficult form of e-communication into a tool for everyone, although it was the mid-1900s before telephones became widely affordable and common in American households.

As the importance and ubiquity of the phone has shown us, e-connection is way better than no connection, but it's nowhere near as good as the real thing. In some ways it brings out the best in us (more on this a bit later), in many ways it brings out the worst in us (more on that in a moment), but in most ways it's just a pale replica of social life in the real world. When people interact face-to-face, they *resonate* with each other, creating interpersonal synchrony in bodily movements, facial expressions, and even pupil dilation and neural activation. This synchrony is already evident in infancy and is influenced by physical touch and eye contact. Synchrony helps people understand each other and it creates rapport. Perhaps e-connection will eventually lead to interpersonal synchrony when our tech and bandwidth are better, but for now there are several ways that it falls down on the job.

First, it turns out that your brain doesn't respond in the same manner when you view someone on the screen as it does when you view them in real life. In a clever demonstration of this effect, Nan Zhao and his colleagues had people engage in conversations across a table from one another. In the 3D condition, people were separated by a glass window in the middle of the table. In the 2D condition, the window was replaced by back-to-back computer screens that showed their conversational partner. Although they still sat across from one another in the 2D condition, they could only see each other on the monitors.

Zhao and colleagues found that people were more engaged in the 3D than the 2D version of the conversation—as indicated by increased pupil dilation and the length of time they looked at each other's faces. Zhao also found greater neural synchrony in the 3D than the 2D condition. These data suggest that even when all other aspects of the conversation are kept constant, e-connection still disrupts interpersonal engagement and synchrony by presenting your face on a flat screen.

Setting aside these effects of flattening your face, other aspects of the conversation are never the same when we e-connect. Most notably, e-connections introduce a brief lag between when you say something and when your partner hears it. This lag can be measured in milliseconds, which might seem trivial, but it's not. When people "click" with each other in a conversation they respond more rapidly than when they don't click. In fact, people who click respond so rapidly—in less than a quarter of a second—that their reactions to each other are largely outside of conscious control. Most forms of e-connection disrupt this sort of rapid responding, as the slight and varying lag times cause people to accidentally interrupt each other when they respond too quickly. As a result, people are forced to put the brakes on when they chat over the internet, making it more difficult to click than when you're socializing in person. Even when you do e-click, people often feel it's not the same because the usual cue—rapid response speed—is missing from the conversation.

Compounding these first two problems, eye contact plays a critical role in interpersonal synchrony but is disrupted by e-connection. Your brain responds differently when someone is looking you in the eyes than when they're not and your pupils come in and out of synchrony as eye contact ebbs and flows throughout the conversation. Eye contact itself is a delicate dance, with episodes of eye contact during a conversation averaging slightly less than two seconds each. In many forms of e-connection the other person's face isn't large enough to know if they're looking you in the eyes and we tend to look at the faces on our screen rather than directly into the camera. Looking at faces gives us the feeling that we're providing eye contact, but looking at the camera gives our conversation partner the feeling that they're receiving eye contact. In combination with a flat face and laggy internet, these technological failings on the part of e-connection result in social interaction that doesn't feel like the real thing.

Compounding this "thin gruel" effect of e-connection, different social media platforms have different vices. First, comparison is the thief of joy and Facebook, Instagram, and the like might as well have been engineered to create the worst sort of social comparisons.

As we discussed in Chapter 5, people curate their posts on social media to emphasize the good and skip over the boring and bad. We know this is true, in part because we do it ourselves, but nonetheless we can't help but feel that everyone else's life on Facebook is better than our own. This social comparison is part of the reason why scrolling through social media often brings our mood down rather than up.

Unfortunately, it's awfully hard to put the brakes on this tendency, largely because social comparison helped our ancestors evaluate their prospects. Our potential ancestors who didn't care how they stacked up to others didn't become our actual ancestors because those who cared worked harder and eventually left the noncaring folks behind. As a result, the tendency to compare ourselves to others is baked into our DNA. Despite our best efforts not to be envious when we see our friends doing great things, it's hard to quash our tendency to engage in this sort of social comparison.

Second, when was the last time you saw people lay into each other in real life like they regularly do on social media? Not all forms of social media are equally guilty in this regard, but I'm astounded by the casual cruelty with which people treat each other on X (the platform formerly known as Twitter). The social distance of e-connection shields us from the direct consequences of our actions, with the result that people are often unpleasant to each other on social media in ways they would never be in real life. When you make fun of someone to their face, you immediately see the sadness or anger in their eyes and can't help but put the brakes on. When you make fun of someone on social media, you're largely blinded to their response.

If you recall the studies by Batson, Cialdini, and their colleagues from Chapter 7, you'll remember that empathy is such a strong emotion that people happily shoulder the suffering of a stranger. But empathy is induced primarily when you confront someone's suffering directly, and social media platforms shield you from most of the suffering experienced by the targets of your ire. As a result, people can be remarkably nasty to each other on social media, particularly when they are tweeting across group boundaries and feel little compunction in expressing their aggression.

By way of example, I recall people saying some rather extraordinary things about me in the online comments of a podcast I did when *The Social Leap* was published. Much of what they wrote was in good humor even if it wasn't exactly complimentary (see comments below), but people often became aggressive and insulting when they disagreed with me, particularly when it came to emotional issues like politics. There were too many comments to wade through all of them, but when I wrote to some of the more aggressive commentators to discuss our disagreements, all but one responded with a friendly note and an apology for the tone in their original comments.

P.: "This guy looks very aerodynamic"

J.: "This guy looks like Jeff Dunham's puppet Peanut."

SP.: "dude looks like a hairless cat"

JB.: "the last airbender has come a long way"

S.: "If Jar Jar Binks & CP30 had a baby it would look like him."

Online commentators discuss my appearance after a podcast.

Online aggression matters, but perhaps the greatest vice of social media platforms is that they encourage us to be lazy in our social habits. Why bother trudging through the snow or taking a subway to a party when you can interact online? Why bother getting together with people for coffee or drinks when you can lie on the couch and FaceTime with them? This laziness in our social habits is particularly pronounced when the outcome of the social event is uncertain, such as when you don't know the people well or are unsure who will be at the party. In such cases, the ease of social media can swamp the uncertain rewards of your social efforts. But some of our best times come when we least expect them, such as when we meet someone new and exciting, and "social media–induced laziness in our social habits," or SMILSH,[*] can prevent that from ever happening.

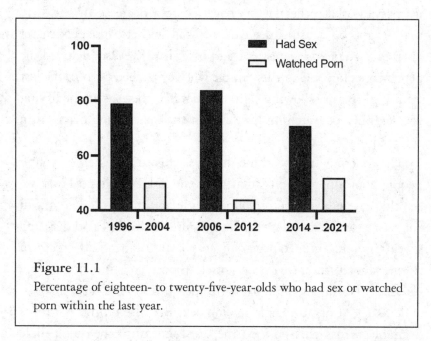

Figure 11.1
Percentage of eighteen- to twenty-five-year-olds who had sex or watched porn within the last year.

There's a fair bit of evidence for SMILSH, but I think the most remarkable example can be seen in changes in sexual activity of young adults since the advent of social media. Given the strength of the human sex drive, some of the most impressive evidence we could find for SMILSH would be if young single adults are having less sex in the era of social media than prior to it. To provide an initial test of the SMILSH hypothesis, we can look at the General Social Survey to estimate the percentage of unmarried people aged eighteen to twenty-five who've had at least one sexual partner in the last year. I've collapsed the data into three slices of

[*] I'm confident this acronym will never catch on, but I can't help myself.

time: 1996–2004 (which is pre–social media), 2006–2012 (which is when social media began to take off), and 2014–2021 (by which time social media penetration in this age group had stabilized at 84 percent).

Because sperm counts have gone down over these time periods, it's possible that other factors have led to a decrease in sex drive that might be causing decreased sexual activity independent of SMILSH. To provide a test of that possibility, I've also included the percentage of people in this sample who've watched porn in the last year. If people are showing signs of SMILSH, they should be having less sex but not watching less porn in the final period depicted in this graph.

The first wave of social media use seems to have helped young people hook up, but consistent with the SMILSH hypothesis, we see that the percentage of young single adults who've had sex in the past year decreases rather sharply in the third period depicted in this graph. Importantly, we see no simultaneous decrease in porn usage. Rather, we see an uptick in porn consumption during the era of social media, although not quite of the same magnitude as the drop in people having sex. Because online porn was available across this entire time, it seems unlikely that people suddenly discovered porn in the third period and decided not to bother having real sex anymore. But it does seem possible that decreased sexual activity is leading to increased porn use, in much the same way that e-connection can be a surrogate for the real thing.

These data provide only the broadest of brushstrokes and there could be numerous other causes of these effects. Nonetheless, the data are consistent with the possibility that social media induces laziness in people's social habits leads them to miss out on interesting opportunities they would otherwise experience. Future research will no doubt provide more compelling evidence one way or the other.

Leveraging Social Media to Reconnect

We're not going to recapture a past in which people joined service clubs and attended weekly meetings (perhaps wearing a funny cap). There's no interest in that possibility and frankly there's no reason to try. The rapid and widespread adoption of social media over the last few decades is essentially a mirror image of the more long-term decline in civic organizations. If we accept that most people are not going to attend local meetings at the Elks Lodge, the ubiquity of e-connection leads us to ask how we might leverage the strengths of social media to reconnect. The starting point in this conversation is to consider the strengths of e-connection and how we might change the way we use social media to maximize the benefits. The positive aspects of e-connection are actually numerous, so let's spend a moment on what I see as the big five.

When used asynchronously, social media gives you more time than face-to-face interactions to formulate a response, providing you with a better chance to say exactly what you mean. One of my college roommates recently called me on Zoom to tell me he's gay. When I asked him how on Earth he was just figuring this out in his late fifties, he said that everyone was so homophobic when we were younger that he was unwilling to accept the possibility that he wasn't straight. That made perfect sense to me, as most of my gay friends didn't start telling me until the nineties, by which time attitudes toward homosexuality had softened a fair bit.

Imagine, however, that he had decided to tell me he's gay when we first started rooming together in the early eighties. I'd like to think I'd have been cool with it and told him that it didn't make any difference to me. But even if I had responded in such a manner, I know my first reaction would have been shock. I probably would have stood there gaping at him for a while, not knowing what to say as the moment became increasingly awkward. And when I did say something, it's easy to imagine that it wouldn't come out right.

In contrast, had he told me over email or some other form of asynchronous social media, I'd have had all the time in the world to get over my surprise and let him know that it didn't make any difference to me. And I'd have also had all the time in the world to get the message just right before sending it. I might have first responded with "I don't care," but on reflection realized that it's a pretty big deal to come out as gay in the eighties, and my reaction might seem *uncaring* rather than supportive. So perhaps I'd have rewritten it, maybe even five or six times, until I had crafted something along the lines of "I'm glad you felt comfortable enough to tell me, but I hope you know that your being gay or straight doesn't change anything about our friendship."

Getting it right isn't easy in the real world, particularly given that conversations are often rapid-fire, we don't choose our words as carefully as we might when we're surprised or upset, and we're often so distracted by concerns about how people see us that we don't pay enough attention to what they actually say. For these reasons, we often wish we could go back and say things better (at least I do). The asynchronous versions of social media allow us to do just that. This feature of social media is so important that people who meet on social media sometimes like each other more than people who meet in real life. This positive effect of social media even emerges when researchers arrange for the same people to (unknowingly) meet in person and online.

As I mentioned in Chapter 8, social media can help you realize that you're not alone after all. No matter how unusual your hobbies, proclivities, or identities, there are others on this planet who share them with you. But those others can be hard to find if your traits are rare, particularly if you don't live in a big city. Social media helps you meet and befriend people just like you no matter how far away they are. Whether your musical, artistic, philosophical, or anime soulmate lives halfway across the planet or just down the road, you're likely to find each other on the same subreddit. We're so accustomed to this aspect of the internet that we forget how isolating it is when people can't find like-minded others. The reassuring effects of social media in this regard can't be overemphasized.

The greater social distance of e-connection facilitates self-disclosure. If I've only recently met you, I might feel awkward talking about issues that go much beyond the weather, recent news, and so on. The social norms encouraging us to start our relationship in the shallow end make perfect sense—with self-disclosure comes vulnerability. But with self-disclosure also comes intimacy and the potential for social support, meaning that our reluctance to self-disclose delays the formation of real friendship. The social distance created by social media reduces the embarrassment of self-disclosure, leading people to disclose earlier and more frequently in their relationships. Increased self-disclosure, in turn, brings people's relationship to the next level faster on social media than in real life, which is one of the (many) reasons why romantic couples are increasingly forming online rather than in person.

Social media platforms enable regular, easy, cheap, and meaningful contact with people regardless of distance. So long as you have an internet connection, social media platforms are essentially free to the user and many include real-time video communication. That allows people to remain part of each other's lives even when circumstances have conspired to pull them apart. For example, when I went to college I chose a university that was thousands of miles from my hometown. Calling home was expensive, so I typically did so only a few times a month, at which point I'd do my best to catch up with my family. We'd talk about the major issues we'd encountered in the last few weeks, but very little came up about the day-to-day small details of our lives.

In the last few years, my daughter has also chosen to go to college thousands of miles from home. She uses a variety of social media platforms that allow her to get in touch on audio and video for free, so she often calls just for a few minutes to get my opinion on random small decisions or tell me about the minor kerfuffles and victories we all encounter in everyday life. Although I still catch up with her on occasional long calls, these random brief video calls allow me to be part of her daily existence in a manner that wasn't possible in the past.

Perhaps the most obvious advantage of social media is that it lets you get together with friends in real time when you can't be there in person. Seeing your friends who are scattered across the globe and talking to them all at once is a whole lot better than not, even if the experience isn't the same as real life. My college roommates and I get together to chat a few times a year over Zoom and it's always great to catch up with the gang. Such platforms are also useful for meeting new people, starting new relationships, joining new clubs, etc.

In circumstances such as these, social media platforms have enormous potential to improve on real life in meaningful ways (even if they'll never be quite as good as real life in other ways). All you need to do is consider the things you don't like about real parties or conferences and how you might fix them online. Here's my top-three list where I think social media already improves (or could easily be tweaked to improve) on real-life social interaction.

1. One of the most embarrassing aspects of face-to-face conversations for many of us is our inability to remember people's names, occupations, or whatever else it is we're meant to know just a few minutes after they told us. Social media largely solves that problem, typically by including people's names underneath their faces when they're chatting with you. The fact that you can't make proper eye contact on video calls is an advantage in this regard, as people can't tell when you stop looking

them in the eye for a moment to remind yourself what their name is. The only analogous situation in real life is when you're at a professional event and everyone is wearing a nametag. But even then you can't possibly look at the nametag on their chest while they're looking at you, so you have to wait for moments when they look away to catch a surreptitious glance and remind yourself of their name.

Videoconferencing platforms have the potential to tell you much more about a person when you're meeting for the first time. For example, I'd like to be able to hover my mouse over people's faces and learn something about them beyond their name, like their occupation, favorite hobbies, where they grew up, or whatever else they'd like people to know. Online platforms can easily provide us with such points of potential connection when we're e-meeting people for the first time (or the second time and have forgotten who they are).

2. When I'm at a party or conference where I don't know many people, there's a lot of potential for awkward moments between conversations. What do I do when I finish one conversation and I'm not talking to anyone else yet? How do I know which conversation would be good to join without blatantly eavesdropping? Who might consider my intrusion into their conversation annoying and who might welcome it? Social media platforms could solve all these problems. First, they could make you invisible at your request, so others don't know you're (still) at the party or conference until you're ready. Second, they could allow people to mark their conversations as private, hide them, or lock them to outsiders when they don't want others to join. Third, when people would welcome others to join their conversation, the platform could allow others to hover their mouse over the conversation to hear what's going on before deciding whether to jump in.

3. In group conversations it would often be handy to be able to communicate a private message to another member of the group without others knowing the content or that a message was even sent. For example, I might want to privately reassure someone else in the conversation that a story is being exaggerated or that the storyteller has some potentially upsetting details wrong. That can be tricky in real life, but is dead easy online (for example, via the chat function on Zoom, which allows you to talk to the entire group or just one member).

Big social media platforms often put their bottom line ahead of the user's experience, typically by developing algorithms that pull you in rather than letting you go when it's time to move on and by creating environments that ramp up hostility rather than kindness. Of course, platforms that don't serve their users' needs create space for platforms that do.

If you live in a highly mobile society, like much of the US and Europe, you probably have good friends and family who've moved hundreds if not thousands of miles away from you (or you from them). The phone is a great tool for keeping up with these people and social media's versatility makes it even better. It's easier than ever to start a relationship on social media and it's also a breeze to keep relationships going through e-contact. As we saw earlier in this chapter, you need to be wary of SMILSH, lest you sabotage your face-to-face contact by replacing it with e-contact when you don't have to. But when your family and friends live a long way away, e-contact is much better than nothing. Just be sure that e-contact remains only a (smallish) part of your social life, as in-person socializing is necessary for good mental health.

If you're the kind of person who joins local civic organizations and gets together with your fellow members in person every week, good on you and keep it up. But, statistically speaking, you're not.

So the key is to follow the Delphic maxim to know yourself and engage in the kind of socializing you can maintain. For most of us, that's a blend of in-person contact with people who live nearby and e-contact with people who don't. In the next chapter, I discuss strategies for increasing both kinds of contact without making your life any busier or more complicated than it is now.

12

Balancing and Rebalancing Your Life

Connection and autonomy are our two most important psychological needs because they played the largest role in helping our ancestors survive and thrive. These needs remain important today but the way we balance them no longer makes sense. As I've argued in Part 3 of this book, people who live in industrialized societies now have an excess of autonomy and not enough connection. This imbalance is most evident among wealthy, educated Western urbanites, but it's common everywhere. If we evolved to be happiest when we balance connection and autonomy in a manner that made our ancestors successful, maybe the way we achieve that balance is no longer ideal. Given how much our society has changed, it's possible the strategies that worked back then are causing us to suffer now. With that possibility in mind, let's start by asking why people keep choosing autonomy over connection.

Why the Relentless Favoring of Autonomy?

Cultural change can be fast or slow. For example, the shift from hunter-gathering to farming took thousands of years, as mobile hunter-gatherers became sedentary farmers over many generations, increasingly supplementing their diet with grains until they eventually planted their own gardens and put down roots. In contrast, the transition from monarchy to democracy in the United States was accomplished in just a few decades, the internet destroyed some industries and created new ones in just a few years, and as of this writing, AI is poised to yield even larger scale and more rapid changes.

Evolutionary change also varies in pace but it's always comparatively slow, relying on differential reproduction and survival rates across numerous generations. Humans evolved to be cultural animals, meaning that one of our greatest strengths is our capacity to leverage cumulative cultural knowledge to solve our own particular problems. The much faster pace of cultural change than biological evolution allows our species to pivot rapidly in response to changing pressures or opportunities. But it also has the potential to result in a mismatch when our cultural practices are at odds with our evolved nature. This mismatch can emerge whenever cultural change is sufficiently rapid that our evolved psychology can't keep pace with newly emerging costs and benefits. In such cases, our evolved proclivities are a poor fit to our daily lives, potentially leading to *miswanting* (wanting things that aren't good for us) or *misfeeling* (feeling emotions that don't reflect the true risks and opportunities facing us).

To get a sense of how such mismatches work and their consequences, let's consider two of the more prominent ones:

1. For almost our entire evolutionary history, we struggled to get enough to eat. In response to this threat, humans evolved a strong desire for salt, fat, and sugar, due to their relative scarcity on the savannah and their importance for our survival. Now that most people in industrialized societies have easy access to as much food as they want, our yearning for salt, fat, and sugar has created an obesity crisis. Too much food is a problem that our ancestors never faced, so we haven't evolved

a propensity to stop wanting salt, fat, and sugar when they're plentiful. This mismatch between our evolved preferences and our new reality puts us at risk, as it is difficult to stop ourselves from consuming excess calories when they're available: a clear example of *miswanting*.

2. Snakes and spiders are funny creatures, in that they don't look as scary as lions and tigers but some of them are just as deadly. Evolution has ensured that we avoid snakes and spiders despite their innocuous appearance by preparing us to fear them; many people are snake or spider phobic who have never suffered from their bite. How deadly are snakes and spiders in our modern world? Not very. The most dangerous spider in the world is the funnel web spider, which lives on the southeast coast of Aus-

Figure 12.1
The world's deadliest spider, moments after hopping out of our groceries.*

---

* And moments before we released it into the yard next to our barbecue! The biologists I was visiting were aghast when I suggested that someone (i.e., one of them) should stomp on it.

tralia. Even though they're common (this one was in a box of produce we bought at the grocery store), the last person to die from a funnel web spider bite was in 1979, because people are good at running away when they encounter them and hospitals keep anti-venom on hand just in case.

Cars and electrical outlets are funny inventions in that they don't look as scary as lions and tigers, but they can kill you just as easily. Every year, nearly fifty thousand people die in motor vehicle accidents in the US alone. Only a few hundred people die from household electric shocks in the US each year, but tens of thousands are accidentally and painfully shocked. Despite the danger that cars and electrical outlets pose, people do not learn to fear them easily; almost no one is car or outlet phobic. Because snakes and spiders posed a recurrent problem throughout our evolutionary history, we've evolved a tendency to fear them even though they're no longer a significant danger. I had trouble relaxing after the funnel web spider popped out of our groceries and everything that brushed against my skin that night gave me the willies. Cars and outlets are evolutionarily novel so we have no evolved response to them whatsoever, despite the ongoing danger they pose. I was once thrown off a ladder by the force of an electric shock, but outlets never give me the heebie-jeebies like spiders do. And I have no trouble relaxing as cars zing by me when I'm out walking, though even the littlest one could easily flatten me, and I know full well that drivers are often distracted by their Big Macs and iPhones. Our tendency to fear snakes and spiders but not cars and electrical outlets is an evolutionary mismatch: a clear example of *misfeeling*.

Can evolutionary mismatches explain why we keep choosing autonomy over connection? Can they account for why populations are migrating to the city when cities make them unhappy? Can they account for why we spend so much time unhappily alone? In short, can they help us understand the numerous situations in which we relentlessly favor autonomy over connection? I think so.

Throughout our evolutionary history, connection was of paramount importance. We needed to form tight connections to survive and that fact was so obvious that we connected unquestioningly. Those hunter-gatherers who couldn't see the need for connection soon became lion chow. As a result, their tendency to go it alone was removed from the gene pool and they also served as a vivid reminder to the folks back home that survival requires connection. For that reason, our genes pushed us to connect, our cultural rules demanded connection, our parental socialization reinforced the message, and our daily lives reminded us that we couldn't live without it. Connection was an obvious necessity.

But autonomy was a luxury. In principle we had an opportunity for autonomy every time our group made a decision we didn't like, but in reality we couldn't just do our own thing. When it came time to break camp, if everyone else wanted to go north and you wanted to go south, you almost always went north. Similar pressures toward conformity were felt when we decided where to hunt, gather raw materials for tools, etc. We either had to persuade people to our point of view or we had to go along with theirs. Life for our hunter-gatherer ancestors was all about connection and compromise, with opportunities for true autonomy popping up only on occasion.

Given this state of affairs, it's likely that we evolved a tendency to grab autonomy whenever it was genuinely available. When autonomy and connection came into conflict, if it was one of those rare circumstances in which we could actually afford to be autonomous, my guess is that's what our ancestors did. According to this possibility, connection was our more important need, but the demands of our daily lives ensured that our connection needs were always met. As a consequence, we evolved to choose autonomy whenever we could get away with it.

In our ancestral past, such situations were rare enough that our default tendency to grab autonomy whenever possible resulted in a proper balance between autonomy and connection. In our modern world, opportunities for autonomy are like salt, fat, and sugar—they're everywhere. As a consequence, our evolved tendency to pick autonomy whenever we're given a genuine choice has become a form of miswanting that has seriously disrupted the balance between these two needs. Where once we were super fit and highly connected because we spent our lives hungry and threatened, now we are out of shape and autonomous because we live in comfort and safety. But just because our modern world allows us to live a certain way doesn't mean that it makes us happy. We may get what we want when we prioritize autonomy, *but we don't get what we need.*

The demands of everyday life no longer force us to maintain connections, but we evolved to connect. As US surgeon general Vivek Murthy argued in his book *Together*, our mental and physical health depend on it. This chapter describes how you can reforge your connections with suggestions that are based on what works. But first let's take a brief detour into the science of self-change to get a sense of what we're up against when we try to change ourselves.

Give It a Try, but Cut Yourself a Break

One of the reassuring things about the field of behavioral genetics is that it teaches us that most human traits are about 50 percent genetic. I take two comforts from that fact. First, although we may not be completely happy with who we are, we are not entirely to blame for our failings. Which sperm met which egg plays as much of a role in our life as anything we've done since. Second, although it's difficult to change, it's possible. My favorite evidence for the fact that our genes make change difficult but not impossible can be seen in Figure 12.2 regarding the genetics of body weight.

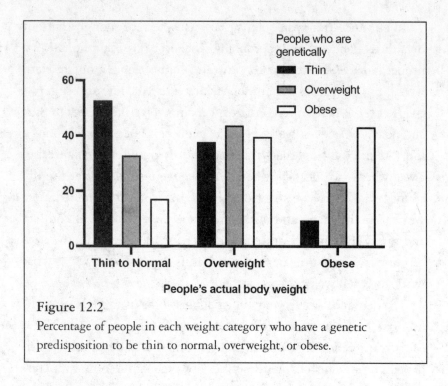

Figure 12.2

Percentage of people in each weight category who have a genetic predisposition to be thin to normal, overweight, or obese.

There are a few things you can see in this graph. First, the data speak to the power of genes. People with thin genes are most likely to be thin to normal weight, people with overweight genes are most likely to be overweight, and people with obese genes are most likely to be overweight or obese. But these data also show us the power of self-determination. Only 9 percent of the people with thin genes are obese, but 17 percent of the people with obese genes are normal weight. Similarly, only 23 percent of the people with overweight genes are obese, but 33 percent of them are normal weight. These data demonstrate that some people need to work harder than others to be who they want to be, but they often succeed, even when the odds are stacked against them. Being normal weight is comparatively easy for people who won this particular genetic lottery and have thin genes, but lots of normal-weight people have overweight or obese genes. Determination matters.

What do these data tell us about the science of self-change more broadly? First and foremost, don't beat yourself up when you fail to make a desired change in your life, as you may be fighting against your genetic makeup. But don't give up either. If one strategy doesn't work, maybe a different one will. Others have fought their genes and won, and you can too. The trick is to choose the right path for you, which may well be the path of least genetic resistance.

Many of the rules that govern your body also govern your mind and what holds true for body weight also holds true for autonomy and connection. If we consider the origins of connection, we see that necessity drew people together when they had to cooperate or perish. As a result, connection is our most fundamental psychological need. But at the same time, it's clear that the extensive connections required by our ancestors' lives necessitated enormous sacrifice. The result was excessive criticism from self and other, endless and psychologically costly social comparisons, and a world in which we had so little autonomy there was almost nothing we could call our own. These costs of tight connections have led people the world over to move toward autonomy when connection is no longer necessary, but by now it should be clear that autonomy without connection is unsatisfying. It should also be clear that too much autonomy and not enough connection describes the lives of most people who live in modern industrialized economies. Particularly if you are well educated, live in the city of a Western country, and have plenty of money, you could probably use more connection in your life.

Five Ways to Restore Connection

So, without further ado . . . let's figure out the best way for you to rebalance. Trust me when I say it's easier than you think, and more rewarding too. Here are the keys to successful life change, starting with the one critical principle from which all else follows.

1. **Change and maintenance must be easy.** As of this writing, more than 40 percent of Americans are obese despite spending over 135 billion dollars each year trying to lose weight. If there's one thing we can learn from these two numbers it's that change is unlikely when it's difficult, no matter how motivated you are. With that concern in the front of our mind, how do we facilitate connection? The key is to incorporate connection into your life in ways that induce the least friction. I started doing this myself with the onset of the COVID pandemic and found it's been easy to maintain because my new protocol is no harder than my old one and a lot more fun. I simply looked at all my activities and for each one that was conducted alone, I asked myself how I might make it more social. Here's my list:

1. I like to do the *New York Times* crossword puzzle and have slowly improved to the point that Monday is easy, Tuesday and Wednesday are challenging, and by Thursday it gets genuinely hard (the *NYT* crossword is designed to get increasingly difficult across the course of the week). My sister lives in London, and I know she likes to do the crossword too, so we started doing them together. She calls once she's made her morning coffee Thursday through Sunday morning (which is late afternoon here in Oz), I put in my earbuds, and we chitchat as we do the puzzle together. I love my sister dearly, but we used to talk only a few times a month because she lives so far away and we're both so busy. The beauty of our joint-puzzle party is that it doesn't add any time to our day; we do something together that we used to do alone.

2. I also enjoy doing the Wordle puzzles, but they're too quick to do them together. Nonetheless, it's fun to see how your friends and family do, so we've created a family Wordle group that

spreads across three generations on WhatsApp, where we share our results each day. Sometimes no one says much of anything other than posting their game, but often we follow it up with comments, and such.*

3. I have a twenty-minute commute in my car a few times a week, which I previously spent listening to music or podcasts. Nothing wrong with music or podcasts, but now if there's anyone I haven't caught up with lately, I call them when I get in the car. My default thought as I get into my car has shifted from "what do I want to hear" to "who would I like to talk to?" It doesn't need to be a car ride either, any mindless alone time will do. My little brother often calls when he's out on his daily run (which I find impressive, as I'm too winded to chat on the rare occasions I find myself running) and my daughter calls when she's doing the laundry or tidying up. You obviously need to keep in mind that the person you're calling might not have time right then, but with some judicious planning you can make sure you talk to people when both of you are busy with a mindless task.

4. My oldest friends in the world are Sid and Richard. We met in the 1960s and have been friends ever since, despite now living thousands of miles apart. Recently we set up a text group to write each other whenever something noteworthy occurs to any of the three of us (with a very loose definition of what it means to be noteworthy). We bounce ideas and random musings off each other relatively often, and then go quiet for weeks at a time. We've known each other so long that we always pick up

---

* For example, my nephew and I are convinced that my cousin is cheating and we're secretly plotting how we're going to catch him red-handed.

where we left off, and the conversations shrink or grow depending on time and interest. Because it's a text chat, we just
pipe in whenever we have time, so our conversations fit around
the rest of our schedule rather than vice versa.

These four fixes might not seem like much, but that's the whole
point. They're small changes in my daily existence that ramp up
my connections with virtually no effort or burden. Let's keep this
basic principle and these examples in mind as we consider what you
might do to change your life.

First and foremost, get together with people in person when
it's relatively easy. That means you should see people face-to-face
when they live nearby and you share joint interests. But it also
means that you should use e-connections to get together virtually
when important people in your life don't live nearby and you share
mutual interests. The mutual interests are the key. In our busy
lives it can be hard to get together with people just to chat, as
we don't have a ton of time for that. But it's easy to get together
with people to do things you want to do anyway because you're
adding connection to existing activities rather than taking time
just to connect. And when you connect to do what you want to do
anyway, you get to have your cake and eat it too; connection and
autonomy are aligned.

The flip side is that you need to be mindful of the fact that
you'll stop getting together with people when it's no longer easy,
even if you still value your relationship. I remember a friend telling
me how one of her closest friends had moved an hour away, but that
they still planned to see each other every weekend. That almost
never happened. A few years later the same friend moved back to
town and they started getting together regularly again, as they're
still very fond of each other. An hour's car ride each way was a
bridge too far for two busy people, despite their mutual interest in

seeing each other. So be realistic about which connections you can maintain easily and plan accordingly.

2. Interventions need to establish new habits. Change is easy if you follow Principle #1, but it can still dissipate over time. The key is to make your new routine habitual or automatic. If you need to plan every time you want to socialize, you'll find you only socialize on occasion in your free time. But if you make connecting a habitual part of your routine, it takes no more effort than brushing your teeth. The best way to establish new habits is to form implementation intentions. By this I mean creating plans that are contingent on something else happening, such as: "When my morning coffee is ready, I'll call my brother to do the crossword." Or "When I'm finished eating dinner, I'll call my old high school friend so we can chat on the phone as we both clean up the dishes."

Notice that each of these plans relies on some environmental trigger for the event to take place. When I'm done making coffee. . . . When dinner is over. . . . The beauty of this approach is that once you make the decision in advance to do X when Y happens, *you don't need to decide again* that you're going to do something. It can be hard to engage in a behavior if you need to continually decide whether to do it. But if you just do it naturally after some other event occurs, engaging in the behavior becomes functionally mindless. In the same way that you don't decide to brush your teeth every morning, you don't want to decide whether to call your friend. Rather, you want that decision to have been made long ago and now all you're doing is executing. Different behaviors take different amounts of time to become habitual, so keep a close eye on your calendar for at least a few months to ensure your new socializing habits become ingrained.

3. Use connection interventions to strengthen other lifestyle goals and vice versa. Most of us have lifestyle goals we're struggling to achieve. They might be exercise goals like going for a twice-weekly run, they might be cleanliness goals like picking up the toddlers' toys every afternoon, or they might be self-improvement goals like taking an online course or learning to play an instrument. By blending your connection goals with your lifestyle goals, you can increase the chance that you'll achieve both (once again, fulfilling autonomy and connection needs at the same time). For example, when I was in college, our tae kwon do team was invited to compete in the national championships. The competition is full contact, and I was a very mediocre martial artist, so I knew that if I was going to survive the tournament I'd have to be in much better shape.*

One of my good friends on the team felt the same way, so even though we both hated getting up early and we both hated running, we decided to go for a run on the hillside near campus every Monday, Wednesday, and Friday morning before class. There's zero chance I would have pulled that off on my own, but joining forces with my pal Tiko worked like a charm. First of all, we took turns being the one who woke the other person. This worked well, because when my alarm rang at seven a.m. (which I regarded as early at the time), my first thought was to turn it off and go back to sleep. But my second thought was the enormous satisfaction I would get from pounding on Tiko's door to wake him up.

Second, once we were out on the road running, I'd be damned if I was going to let him beat me (though he often did). That meant we both started strong and as we approached the last part of the run, we were both motivated to begin our final sprint earlier than the other guy to get a bit of a lead. By the time the tournament came around, I was in the best shape of my life, and although my opponent in the first round used me for target practice, he was more winded after beating me up than I was. You may or may not be practicing for a martial arts tournament, but you could incorporate connection into your lifestyle goals by running or biking with

---

* And not just to run away, although the thought did occur to me.

a friend, lifting weights together, hiking together, or gardening or doing other home activities together. Doing such activities together in person is best, but as I said, e-connection is better than no connection at all.

4. **Connect at work.** Don't eat in front of the computer, eat with the person in finance whose advice you need anyway. Or schedule a regular lunch with your mentor. Don't commute to work alone if you can arrange a carpool. You'll be happier and you'll save the planet at the same time. You don't need to do these things every day, but you do need to do them on a regular schedule if you're going to follow Principle #2 and establish new habits. It's not enough to say "let's carpool when it's convenient." You need to sort out the days it'll be convenient for both of you, and then stick with the plan if at all possible. That way the Tuesday carpool becomes a habit and you automatically start spending more time with colleagues. These people don't need to be your best friends, they just need to be better company than an empty room or automobile.

If you work remotely, you can still connect at work if you think outside the box. For example, one of my friends has long held regular one-on-one meetings with people on her team, in which they lay out tasks and plans for the upcoming week. With the onset of the pandemic, these meetings moved to Zoom. The meetings also started becoming more social as people wanted to make up for lost lunches and after-work drinks, so she started scheduling them to start during coffee or lunch breaks whenever possible. With this change, the Zoom calls became e-coffees or e-lunches that segued into a formal meeting once the two of them had caught up about whatever was going on in their lives.

Because e-comms allow easy integration of audio-visuals, she also started thinking of fun things to add to the beginning of each meeting. She's a keen rock climber and so is one of her colleagues, so each week one of them was assigned to bring a video of their coolest climb or most epic fail to start out the meeting, which proved to be great fun. From what I can tell, she and her colleagues enjoyed these connections almost as much as the real thing.

5. Don't do things alone (except when you want to or need to). Take a look at all the activities you do throughout the day and ask yourself how each and every one of them could become more social—all the way down to walking the dog or picking up groceries. Sometimes you won't know anyone who likes to do what you do, but that shouldn't stymie you. If your friends don't jog, join a running club. It might not work out and you might find you're happier running alone, but there's a decent chance you'll make new friends with shared interests when you link your activities with theirs. The same holds for lots of other activities. If you like video games, play them online with other gamers rather than by yourself. If you like to paint, join a studio where other people go to paint. Although this may sound absurd, if you enjoy activities that don't easily integrate with others but can be done in public, start doing them at a local park or other public space. For example, if you're not too easily distracted, try working or reading in a library reading room or coffee shop. The micro-interactions that activities in public create are better than no interactions at all and you never know where they might lead.

Many of these suggestions are easier and more fun if you're an extrovert than if you're an introvert, but that doesn't mean you shouldn't do these things if you're introverted. It just means you should adapt the various strategies to suit your preferences. Introverts need social connection almost as much as extroverts do, they just typically prefer them in smaller doses with fewer people. The first four strategies should work well if you're introverted so long as you limit your socializing to just a few people at a time. But if you're an introvert you might think some of the suggestions in Principle #5 sound aversive. No harm in giving #5 a miss entirely, but there are probably ways you can make such activities fun and workable. The key is to do them in a context where you don't feel awkward if you're not socializing but in which some socializing is also natural. Painting in a studio would be a good example of such a situation, as you aren't required to say a word to the person next to you, but it's easy to stop and admire their painting and strike up a conversation if you want to.

Rebalancing Is Not a One-Off

None of us are the same person we used to be, nor are we the same person we'll be in a decade or two. Our personality doesn't change much but our goals and motives change a lot, so what works for you now won't work for you permanently. When you're younger, it makes sense for you to be an autonomy machine with temporary connections that supplement a few long-term friendships. This is the time in life to develop the skills and gain the experience that'll make you successful, so autonomy is paramount. Your important friends from your past are likely to be pursuing their own goals, so everyone is understanding when you cancel plans or even move away for work. It's also the time in life when you'll meet the most people, with many of the resulting relationships proving to be transient. You may need to kiss a lot of frogs to meet your prince or princess, and once you realize you're snogging a frog, most people move on.

Although this set of life goals and strategies dominates adolescence and early adulthood, depending on your career and dating history it usually fades with your twenties. By the time people are in their thirties, they're often starting a family, meaning that connection reemerges as a key life goal. People are also more secure in their career at this point, with a clearer ladder ahead of them, so autonomy remains important but is no longer paramount given their greater job security. This process continues naturally into middle adulthood, with career and autonomy competing for center stage with family and connection. When we talk about work/life balance, this is usually the stage we're talking about. People are unfazed if you work like a dog in your early twenties as you seek to establish yourself, but it's increasingly problematic if you continue this habit into your thirties and even forties.

By the time people get to their fifties they're often at the peak of their career, which means they're starting to look down into the valley on the other side. As they envision ambling down that hill in their sixties or seventies, they often ask themselves what they'll do for entertainment, having spent so many hours working that it's hard to remember what else a person might do. This is the point at which connection regains center stage and autonomy starts to recede to a few residual domains, such as the freedom to choose your own activities and sources of entertainment. No one wants to be a failure at work, but the idea of aging and dying alone is more than most of us can bear.

This natural progression in the balance between autonomy and connection, with some individual differences and a bit of wobbling along the way, means that rebalancing is not a one-off event. Rather, rebalancing is a long-term strategy that involves continual assessment and small tweaks as needed. If you make minor course corrections along the way, you'll be much less likely to find yourself completely lost when you stop and reflect on your life at your fiftieth birthday. Continual assessment might sound like a nuisance, but it's not something you need to do every morning before breakfast. Rather, it's something that should pop up on your calendar every year or so *and* whenever you're considering major life changes.

To begin with the "every year or so," rebalancing is a lot easier if you don't let your life get completely out of whack. If I realize that I'm letting my friends fall by the wayside before they've forgotten my name, it's a relatively simple matter to get back in touch and reconnect. But if they say "Bill who?" when I call on the phone, I've got my work cut out for me. Additionally, if we fall into a routine that doesn't incorporate our friends, we start getting a bit picky about how we do things. That, in turn, makes it more challenging to bring your friends back into your life due to the inherent tension between autonomy and connection that we've discussed throughout this book. If your friends want to go for a jog in the evening and you want to go in the morning, it's easier to compromise before you get too set in your ways. The longer we've done something a certain way, the more inconvenient it is to change, as the activities in our lives become interlocked with each other, with every potential change requiring additional changes elsewhere.

Moments of major life changes are also a great point for reassessment. If you're thinking of moving to a new town or taking a new job, your life is going to get turned upside down anyway. You're already throwing a spanner in the works, so now is a perfect time to ask yourself if you're happy with your current balance between connection and autonomy. If the answer isn't a resounding yes, then you might check the list of strategies we've just discussed to see which of them you can easily fold into your new existence. Given that your new existence is essentially a blank slate, you have no excuse for not trying a few of the more promising approaches.

In the last two chapters I've offered advice about how to rebalance your life, but you're the only one who can decide which pieces of advice suit you. In psychology there is no one-size-fits-all approach; everything is bespoke. If your life could use a little rebalancing, I'd recommend that you start with a strategy that looks easy and likely to be fun. But don't choose one that would be hard to undo if it proves to be a disaster—once you invite your boss on your daily run, it might be hard to uninvite her if she proves to be more hassle than she's worth. I'd also recommend that you tailor my suggestions to your proclivities. Try one thing, give it a few weeks, and see if it's worth it. Are you happier? Are you enjoying your day more? At the very least, is the change not making matters worse?

The same holds for those of you who received this book from

a friend or family member. Just because others believe your life is unbalanced doesn't mean it is, but it does raise the possibility that the gift giver might be right. Why not choose an easy strategy that doesn't look too bad and give it a try?

It's a remarkable fact that we're no happier than hunter-gatherers, given the incredibly difficult and dangerous lives they lead, but it's true. My goal in writing this book has been to provide a comprehensive theory for how this remarkable state of affairs came to be. There's a lot of great advice out there about how to make yourself happier, and I encourage you to incorporate anything that works for you. But I also encourage you to think about what lies at the base of the problem. If I'm right, and we have shifted to a world that overemphasizes autonomy at a cost to connection, then all the micro-fixes offered by the numerous approaches to happiness are really just Band-Aids.

There's nothing wrong with a Band-Aid, but it won't help you heal unless you also address the ongoing source of your injury. As I've argued throughout this book, the source of your injury is your lack of connection caused by your relentless pursuit of autonomy. As should be evident in this final chapter, once you see the problem, it's not hard to fix. I've laid out a number of approaches in this chapter that are painless and easy. But you'll also need to be mindful as you make future decisions in your life, both big and small, that you're not repeatedly prioritizing autonomy over connection. Choosing autonomy has become a habit for most of us and it takes a fair bit of attention and self-awareness to break bad habits. So you should expect to stumble a few times along the way, but once you start adding connection back into your daily life, that, too, will become automatic and easy. It should also be incredibly rewarding.

Acknowledgments

Once again, I owe a huge thanks to Lauren Sharp, my agent at Aevitas Creative Management, who helped me shape some rather amorphous ideas into this book. She then handed the baton to my wonderful editor, Kirby Sandmeyer, who asked all the right questions and proved once again that the (red) pen is mightier than the sword. And, of course, thanks to Karen Rinaldi, who decided to invest in this book.

My friends and family also played a critical role along the way, reading and commenting on early drafts that weren't ready for public consumption. A huge thanks to Courtney, Maya, Jordy, Ted, Frank, Marianne, Karin, and Paul vH, Meris Van de Grift, Amanda Niehaus, Jade Butterworth, Leina Gries, Joe Forgas, John Merakovsky, Adam Cohen, Ole Höffken (thanks also Ole for inspiring the title), Peter McGraw, Richard Drake, and Lee Prescott. And a particularly big thanks to David Boninger, who wasted a good chunk of his sabbatical reading chapter drafts and discussing them with me over some very tasty mocha coffees.

References

INTRODUCTION:

Marlowe, Frank. *The Hadza: Hunter-gatherers of Tanzania.* Berkeley: University of California Press, 2010.

Poirotte, Clémence, Peter M. Kappeler, Barthelemy Ngoubangoye, Stéphanie Bourgeois, Maick Moussodji, and Marie JE Charpentier. "Morbid attraction to leopard urine in Toxoplasma-infected chimpanzees." *Current Biology* 26, no. 3 (2016): R98–R99.

Stewart, Brian A., Yuchao Zhao, Peter J. Mitchell, Genevieve Dewar, James D. Gleason, and Joel D. Blum. "Ostrich eggshell bead strontium isotopes reveal persistent macroscale social networking across late Quaternary southern Africa." *Proceedings of the National Academy of Sciences* 117, no. 12 (2020): 6453–6462.

Vyas, Ajai, Seon-Kyeong Kim, Nicholas Giacomini, John C. Boothroyd, and Robert M. Sapolsky. "Behavioral changes induced by Toxoplasma infection of rodents are highly specific to aversion of cat odors." *Proceedings of the National Academy of Sciences* 104, no. 15 (2007): 6442–6447.

Wiessner, P. "Risk, reciprocity and social influences on !Kung San economics" in *Politics and History in Band Societies*, E. Leacock, R. B. Lee, eds. Cambridge, UK: Cambridge University Press, 1982, pp. 61–84.

CHAPTER 1:

Arriaga, Ximena B., Madoka Kumashiro, Eli J. Finkel, Laura E. VanderDrift, and Laura B. Luchies. "Filling the void: Bolstering attachment security in committed relationships." *Social Psychological and Personality Science* 5, no. 4 (2014): 398–406.

De Zalduondo, Barbara O. "Ecology and affective behavior: Selected results from a quantitative study among Efe foragers of northeast Zaire." *American Journal of Physical Anthropology* 78, no. 4 (1989): 533–545.

Frackowiak, Tomasz, Anna Oleszkiewicz, Marina Butovskaya, Agata Groyecka, Maciej Karwowski, Marta Kowal, and Piotr Sorokowski. "Subjective happiness among Polish and Hadza people." *Frontiers in Psychology* 11 (2020): 1173.

Haidt, Jonathan. "Social media is a major cause of the mental illness epidemic in teen girls. Here's the evidence." After Babel, February 22, 2023. https://jonathanhaidt.substack.com/p/social-media-mental-illness-epidemic ?ref=quillette.

Haidt, Jonathan. *The anxious generation: How the great rewiring of childhood is causing an epidemic of mental illness.* New York: Random House, 2024.

Haidt, Jonathan, and Greg Lukianoff. *The Coddling of the American Mind: How Good Intentions and Bad Ideas Are Setting Up a Generation for Failure.* London: Penguin, 2018.

Kumashiro, Madoka, Caryl E. Rusbult, and Eli J. Finkel. "Navigating personal and relational concerns: The quest for equilibrium." *Journal of Personality and Social Psychology* 95, no. 1 (2008): 94.

Ryan, Richard M., and Edward L. Deci. *Self-Determination Theory: Basic Psychological Needs in Motivation, Development, and Wellness.* New York: Guilford Press, 2017.

Wang, Yan, and Krishna Savani. "The salience of choice reduces social responsibility: evidence from lab experiments and compliance with COVID-19 stay-at-home orders." *PNAS Nexus* 1, no. 4 (2022): pgac200.

CHAPTER 2:

Bridgland, Victoria M. E., Benjamin W. Bellet, and Melanie K. T. Takarangi. "Curiosity Disturbed the Cat: Instagram's Sensitive-Content Screens Do Not Deter Vulnerable Users From Viewing Distressing Content." *Clinical Psychological Science* 11, no. 2 (2022): 290–307.

Butterworth, J., D. Smerdon, R. Baumeister, and W. von Hippel (in press). "Cooperation in the time of Covid." *Perspectives on Psychological Science.*

Connor, Richard C., Michael Krützen, Simon J. Allen, William B. Sherwin, and Stephanie L. King. "Strategic intergroup alliances increase access to a contested resource in male bottlenose dolphins." *Proceedings of the National Academy of Sciences* 119, no. 36 (2022): e2121723119.

Dean, Lewis G., Rachel L. Kendal, Steven J. Schapiro, Bernard Thierry, and Kevin N. Laland. "Identification of the social and cognitive processes underlying human cumulative culture." *Science* 335, no. 6072 (2012): 1114–1118.

Granovetter, Mark S. "The strength of weak ties." *American Journal of Sociology* 78, no. 6 (1973): 1360–1380.

Kitanishi, Koichi. "Food sharing among the Aka hunter-gatherers in northeastern Congo." *African Study Monographs. Supplementary issue* 25 (1998): 3–32.

Leary, Mark R., and R. F. Baumeister. "The need to belong." *Psychological Bulletin* 117, no. 3 (1995): 497–529.

Marlowe, Frank. *The Hadza: Hunter-gatherers of Tanzania*. Berkeley: University of California Press, 2010.

Marlowe, Frank W. "Dictators and Ultimatums in an Egalitarian Society of Hunter-Gatherers: The Hadza of Tanzania" in *Foundations of Human Sociality*, J. Henrich et al., eds. New York: Oxford University Press, 2004, pp. 168–193.

Rajkumar, Karthik, Guillaume Saint-Jacques, Iavor Bojinov, Erik Brynjolfsson, and Sinan Aral. "A causal test of the strength of weak ties." *Science* 377, no. 6612 (2022): 1304–1310.

Righetti, Francesca, Catrin Finkenauer, and Eli J. Finkel. "Low self-control promotes the willingness to sacrifice in close relationships." *Psychological Science* 24, no. 8 (2013): 1533–1540.

Woodburn, James. "Egalitarian societies revisited." *Property and Equality* 1 (2005): 18–31.

CHAPTER 3:

Ames, Kenneth M. "The northwest coast." *Evolutionary Anthropology: Issues, News, and Reviews* 12, no. 1 (2003): 19–33.

Epley, Nicholas, Amit Kumar, James Dungan, and Margaret Echelbarger. "A prosociality paradox: How miscalibrated social cognition creates a misplaced barrier to prosocial action." *Current Directions in Psychological Science* 32, no. 1 (2022): 33–41.

Kervyn, Nicolas, Vincent Yzerbyt, and Charles M. Judd. "Compensation between warmth and competence: Antecedents and consequences of a negative relation between the two fundamental dimensions of social perception." *European Review of Social Psychology* 21, no. 1 (2010): 155–187.

Klingenstein, Sara, Tim Hitchcock, and Simon DeDeo. "The civilizing process in London's Old Bailey." *Proceedings of the National Academy of Sciences* 111, no. 26 (2014): 9419–9424.

Langer, E. J. "The illusion of control." *Journal of Personality and Social Psychology* 32 (1975): 311–328.

Martins, Mauricio de Jesus Dias, and Nicolas Baumard. "The rise of prosociality in fiction preceded democratic revolutions in Early Modern Europe." *Proceedings of the National Academy of Sciences* 117, no. 46 (2020): 28684–28691.

Pinker, Steven. *The Better Angels of Our Nature: The Decline of Violence in History and Its Causes*. London: Penguin, 2011.

———. *Enlightenment Now: The Case for Reason, Science, Humanism, and Progress*. London: Penguin, 2018.

Suddendorf, Thomas, Jonathan Redshaw, and Adam Bulley. *The Invention of Tomorrow: A Natural History of Foresight*. New York: Basic Books, 2022.

Warren, Samuel D., and Louis D. Brandeis (1890). "The Right to Privacy." *Harvard Law Review* 4, no. 5 (1890): 193–220.

"What is privacy." Office of the Australian Information Commissioner, Australian Government. https://www.oaic.gov.au/privacy/your-privacy-rights/your-personal-information/what-is-privacy.

CHAPTER 4:

Bertrand, Marianne, and Emir Kamenica. "Coming apart? Cultural distances in the United States over time." Working paper 24771. National Bureau of Economic Research, 2018.

Bruce, Vicki, A. Mike Burton, Elias Hanna, Pat Healey, Oli Mason, Anne Coombes, Rick Fright, and Alf Linney. "Sex discrimination: How do we tell the difference between male and female faces?" *Perception* 22, no. 2 (1993): 131–152.

Cameron, Catherine M. *Captives: How Stolen People Changed the World*. Lincoln: University of Nebraska Press, 2016.

Ceci, Stephen J., Donna K. Ginther, Shulamit Kahn, and Wendy M. Williams. "Women in academic science: A changing landscape." *Psychological Science in the Public Interest* 15, no. 3 (2014): 75–141.

Cibis, Anna, Roland Mergl, Anke Bramesfeld, David Althaus, Günter Niklewski, Armin Schmidtke, and Ulrich Hegerl. "Preference of lethal methods is not the only cause for higher suicide rates in males." *Journal of Affective Disorders* 136, no. 1–2 (2012): 9–16.

Cox, Daniel A. "The state of American friendship: Change, challenges, and loss." Survey Center on American Life, June 8, 2021. https://www.americansurveycenter.org/research/the-state-of-american-friendship-change-challenges-and-loss/.

Falk, Armin, and Johannes Hermle. "Relationship of gender differences in preferences to economic development and gender equality." *Science* 362, no. 6412 (2018): eaas9899.

Giurge, Laura M., Ashley V. Whillans, and Ayse Yemiscigil. "A multicountry perspective on gender differences in time use during COVID-19." *Proceedings of the National Academy of Sciences* 118, no. 12 (2021): e2018494118.

Gough, Christina. "NFL in the United States as of April 2023, by gender." Statista, June 4, 2024. https://www.statista.com/statistics/1098882/interest-level-football-gender/.

Hooven, Carole. *Testosterone: The Story of the Hormone that Dominates and Divides Us*. Hachette UK, 2021.

Hrdy, Sarah Blaffer. *Mothers and Others: The Evolutionary Origins of Mutual Understanding*. Cambridge, MA: Harvard University Press, 2009.

Jelenkovic, A., R. Sund, Y. M. Hur et al. "Genetic and environmental influences on height from infancy to early adulthood: An individual-based pooled analysis of 45 twin cohorts." *Scientific Reports* 6, 28496 (2016). https://doi.org/10.1038/srep28496.

Lubinski, David, Camilla P. Benbow, Kira O. McCabe, and Brian O. Bernstein. "Composing Meaningful Lives: Exceptional Women and Men at Age 50." *Gifted Child Quarterly* 67, no. 4 (2023): 278–305.

Midgette, Allegra J., Danyang Ma, Lucy M. Stowe, and Nadia Chernyak. "US and Chinese preschoolers normalize household labor inequality." *Proceedings of the National Academy of Sciences* 120, no. 38 (2023): e2301781120.

Özçalışkan, Şeyda, and Susan Goldin-Meadow. "Sex differences in language first appear in gesture." *Developmental Science* 13, no. 5 (2010): 752–760.

Salk, Rachel H., Janet S. Hyde, and Lyn Y. Abramson. "Gender differences in depression in representative national samples: Meta-analyses of diagnoses and symptoms." *Psychological Bulletin* 143, no. 8 (2017): 783.

Sher, Leo. "Gender differences in suicidal behavior." *QJM: An International Journal of Medicine* 115, no. 1 (2022): 59–60.

CHAPTER 5:

Diener, Ed, Louis Tay, and David G. Myers. "The religion paradox: If religion makes people happy, why are so many dropping out?" *Journal of Personality and Social Psychology* 101, no. 6 (2011): 1278.

English, Alexander Scott, Thomas Talhelm, Rongtian Tong, Xiaoyuan Li, and Yan Su. "Historical rice farming explains faster mask use during early days of China's COVID-19 outbreak." *Current Research in Ecological and Social Psychology* 3 (2022): 100034.

Gebauer, Jochen E., Jennifer Eck, Theresa M. Entringer, Wiebke Bleidorn, Peter J. Rentfrow, Jeff Potter, and Samuel D. Gosling. "The well-being benefits of person-culture match are contingent on basic personality traits." *Psychological Science* 31, no. 10 (2020): 1283–1293.

Gelfand, Michele J., Jana L. Raver, Lisa Nishii, Lisa M. Leslie, Janetta Lun, Beng Chong Lim, Lili Duan et al. "Differences between tight and loose cultures: A 33-nation study." *Science* 332, no. 6033 (2011): 1100–1104.

Harrington, Jesse R., and Michele J. Gelfand. "Tightness–looseness across the 50 united states." *Proceedings of the National Academy of Sciences* 111, no. 22 (2014): 7990–7995.

Henrich, Joseph. *The WEIRDest People in the World: How the West Became Psychologically Peculiar and Particularly Prosperous.* London: Penguin, 2020.

Hofstede, Geert. *Culture's Consequences: Comparing Values, Behaviors, Institutions and Organizations Across Nations.* Thousand Oaks, CA: Sage Publications, 2001. https://geerthofstede.com.

Kitayama, Shinobu, Hazel Rose Markus, Hisaya Matsumoto, and Vinai Norasakkunkit. "Individual and collective processes in the construction of the self: Self-enhancement in the United States and self-criticism in Japan." *Journal of Personality and Social Psychology* 72, no. 6 (1997): 1245.

Lee, Cheol-Sung, Thomas Talhelm, and Xiawei Dong. "People in historically rice-farming areas are less happy and socially compare more than people in wheat-farming areas." *Journal of Personality and Social Psychology* 124, no. 5 (2022): 935–957.

Melton, R. Jeffrey, and Robert C. Sinclair. "Culture and COVID-19: A global analysis of the successes of collectivist countries and the failures of individualistic countries." *Available at SSRN 3954093* (2021).

Motyl, Matt, Ravi Iyer, Shigehiro Oishi, Sophie Trawalter, and Brian A. Nosek. "How ideological migration geographically segregates groups." *Journal of Experimental Social Psychology* 51 (2014): 1–14.

Okazaki, S. "Sources of ethnic differences between Asian American and White American college students on measures of depression and social anxiety." *Journal of Abnormal Psychology* 106 (1997): 52–60.

Santos, Henri C., Michael E. W. Varnum, and Igor Grossmann. "Global increases in individualism." *Psychological Science* 28, no. 9 (2017): 1228–1239.

Sedikides, Constantine, Lowell Gaertner, and Yoshiyasu Toguchi. "Pancultural self-enhancement." *Journal of Personality and Social Psychology* 84, no. 1 (2003): 60.

Steptoe, Andrew, Jane Ardle, Akira Tsuda, and Yoshiyuki Tanaka. "Depressive symptoms, socio-economic background, sense of control, and cultural factors in university students from 23 countries." *International Journal of Behavioral Medicine* 14 (2007): 97–107.

Talhelm, Thomas, and Alexander S. English. "Historically rice-farming societies have tighter social norms in China and worldwide." *Proceedings of the National Academy of Sciences* 117, no. 33 (2020): 19816–19824.

Talhelm, Thomas, Xiao Zhang, Shige Oishi, Chen Shimin, Dechao Duan, Xiaoli Lan, and Shinobu Kitayama. "Large-scale psychological differences within China explained by rice versus wheat agriculture." *Science* 344, no. 6184 (2014): 603–608.

Thomson, Robert, Masaki Yuki, Thomas Talhelm, Joanna Schug, Mie Kito, Arin H. Ayanian, Julia C. Becker et al. "Relational mobility predicts social behaviors in 39 countries and is tied to historical farming and threat." *Proceedings of the National Academy of Sciences* 115, no. 29 (2018): 7521–7526.

Woody, Sheila R., Sheena Miao, and Kirstie Kellman-McFarlane. "Cultural differences in social anxiety: A meta-analysis of Asian and European heritage samples." *Asian American Journal of Psychology* 6, no. 1 (2015): 47.

Yoo, Jiah, and Yuri Miyamoto. "Cultural fit of emotions and health implications: A psychosocial resources model." *Social and Personality Psychology Compass* 12, no. 2 (2018): e12372.

CHAPTER 6:

"Americans Donated More than $3 Billion to Tsunami Relief Efforts, Study Finds." Philanthropy News Digest, December 26, 2006. https://philanthropy newsdigest.org/news/americans-donated-more-than-3-billion-to-tsunami -relief-efforts-study-finds.

Barrett, H. Clark, Alexander Bolyanatz, Alyssa N. Crittenden, Daniel M. T. Fessler, Simon Fitzpatrick, Michael Gurven, Joseph Henrich et al. "Small-scale societies exhibit fundamental variation in the role of intentions in moral judgment." *Proceedings of the National Academy of Sciences* 113, no. 17 (2016): 4688–4693.

Boehm, C. *Hierarchy in the Forest: The Evolution of Egalitarian Behavior.* Cambridge, MA: Harvard University Press, 2001.

———. *Moral Origins: The Evolution of Virtue, Altruism, and Shame.* New York: Basic Books, 2012.

Boyer, Pascal. *Religion Explained.* New York: Random House, 2008.

Butterworth, J., D. Smerdon, R. Baumeister, and W. von Hippel (in press). "Cooperation in the time of Covid." *Perspectives on Psychological Science.*

Cohen, Adam B. "Religion's profound influences on psychology: Morality, intergroup relations, self-construal, and enculturation." *Current Directions in Psychological Science* 24, no. 1 (2015): 77–82.

Cohen, Adam B., and Peter C. Hill. "Religion as culture: Religious individualism and collectivism among American Catholics, Jews, and Protestants." *Journal of Personality* 75, no. 4 (2007): 709–742.

Cohen, Adam B., and Paul Rozin. "Religion and the morality of mentality." *Journal of Personality and Social Psychology* 81, no. 4 (2001): 697.

Ialongo, Nicola, Raphael Hermann, and Lorenz Rahmstorf. "Bronze Age weight systems as a measure of market integration in Western Eurasia." *Proceedings of the National Academy of Sciences* 118, no. 27 (2021): e2105873118.

Li, Yexin Jessica, Kathryn A. Johnson, Adam B. Cohen, Melissa J. Williams, Eric D. Knowles, and Zhansheng Chen. "Fundamental(ist) attribution error: Protestants are dispositionally focused." *Journal of Personality and Social Psychology* 102, no. 2 (2012): 281.

"A Quarter of Americans Have Donated to Supported [*sic*] Ukraine, Survey Finds." Philanthropy News Digest, March 22, 2022. https://philanthropynews digest.org/news/a-quarter-of-americans-have-donated-to-supported-ukraine -survey-finds.

"Religious landscape study." Pew Research Center, 2014. https://www.pew research.org/religion/religious-landscape-study/party-affiliation/.

Ritter, Ryan S., Jessé Lee Preston, and Ivan Hernandez. "Happy tweets: Christians are happier, more socially connected, and less analytical than atheists on Twitter." *Social Psychological and Personality Science* 5, no. 2 (2014): 243–249.

Smith, Tom W., Michael Davern, Jeremy Freese, and Stephen Morgan. General Social Surveys, 1972–2018 [machine-readable data file] /Principal Investigator, Tom W. Smith; Co-Principal Investigators, Michael Davern, Jeremy Freese, and Stephen Morgan; Sponsored by National Science Foundation. NORC ed. (Chicago: NORC, 2018). NORC at the University of Chicago Data accessed from the GSS Data Explorer website at gssdataexplorer.norc .org.

Starmans, Christina, and Paul Bloom. "When the spirit is willing, but the flesh is weak: Developmental differences in judgments about inner moral conflict." *Psychological Science* 27, no. 11 (2016): 1498–1506.

Vardy, Tom, Cristina Moya, Caitlyn D. Placek, Coren L. Apicella, Alexander Bolyanatz, Emma Cohen, Carla Handley et al. "The religiosity gender gap in 14 diverse societies." *Religion, Brain & Behavior* 12, no. 1–2 (2022): 18–37.

Watts, Joseph, Simon J. Greenhill, Quentin D. Atkinson, Thomas E. Currie, Joseph Bulbulia, and Russell D. Gray. "Broad supernatural punishment but not moralizing high gods precede the evolution of political complexity in Austronesia." *Proceedings of the Royal Society B: Biological Sciences* 282, no. 1804 (2015): 20142556.

CHAPTER 7:

Abramson, Lior, Florina Uzefovsky, Virgilia Toccaceli, and Ariel Knafo-Noam. "The genetic and environmental origins of emotional and cognitive empathy: review and meta-analyses of twin studies." *Neuroscience & Biobehavioral Reviews* 114 (2020): 113–133.

Boaz, David, and David Kirby. "The libertarian vote." *Cato Institute Policy Analysis Series* 580 (2006).

Casey, James P., Eric J. Vanman, and Fiona Kate Barlow. "Empathic conservatives and moralizing liberals: Political intergroup empathy varies by political ideology and is explained by moral judgment." *Personality and Social Psychology Bulletin* (2023): 01461672231198001.

Cialdini, Robert B., Mark Schaller, Donald Houlihan, Kevin Arps, Jim Fultz, and Arthur L. Beaman. "Empathy-based helping: Is it selflessly or selfishly motivated?" *Journal of Personality and Social Psychology* 52, no. 4 (1987): 749.

Delmore, Erin. "This is how women voters decided the 2020 election." MSNBC, November 13, 2020. https://www.msnbc.com/know-your-value /how-women-voters-decided-2020-election-n1247746.

"Facts and Figures: World's Most Venomous Snakes." Australian Venom Research Unit. https://web.archive.org/web/20150111055930/http://www.avru .org/?q=general%2Fgeneral_mostvenom.html.

Fallon, Nicholas, Carl Roberts, and Andrej Stancak. "Shared and distinct functional networks for empathy and pain processing: a systematic review and meta-analysis of fMRI studies." *Social Cognitive and Affective Neuroscience* 15, no. 7 (2020): 709–723.

General Social Survey, NORC, University of Chicago.

Goldberg, J. H., J. S. Lerner, and P. E. Tetlock. "Rage and reason: The psychology of the intuitive prosecutor." *European Journal of Social Psychology* 29 (1999): 781–795.

Graham, Jesse, Jonathan Haidt, and Brian A. Nosek. "Liberals and conservatives rely on different sets of moral foundations." *Journal of Personality and Social Psychology* 96, no. 5 (2009): 1029.

Greenberg, David M., Varun Warrier, Ahmad Abu-Akel, Carrie Allison, Krzysztof Z. Gajos, Katharina Reinecke, P. Jason Rentfrow, Marcin A. Radecki, and Simon Baron-Cohen. "Sex and age differences in 'theory of mind' across 57 countries using the English version of the 'Reading the Mind in the Eyes' Test." *Proceedings of the National Academy of Sciences* 120, no. 1 (2023): e2022385119.

Iyer, Ravi, Spassena Koleva, Jesse Graham, Peter Ditto, and Jonathan Haidt. "Understanding libertarian morality: The psychological dispositions of self-identified libertarians." *PLoS One* 7, no. 8 (2012): e42366.

Johnson, Dominic D. P., Daniel T. Blumstein, James H. Fowler, and Martie G. Haselton. "The evolution of error: Error management, cognitive constraints, and adaptive decision-making biases." *Trends in Ecology & Evolution* 28, no. 8 (2013): 474–481.

Jones, Jeffrey M. "Gender gap in 2012 vote is largest in Gallup's history." Gallup, November 9, 2012. https://news.gallup.com/poll/158588/gender-gap-2012-vote-largest-gallup-history.aspx#:~:text=PRINCETON%2C%20NJ%20%2D%2D%20President%20Barack,over%20Republican%20challenger%20Mitt%20Romney.

Lerner, J. S., J. H. Goldberg, and P. E. Tetlock. "Sober second thought: The effects of accountability, anger and authoritarianism on attributions of responsibility." *Personality and Social Psychology Bulletin* 24 (1998): 563–574.

McCue, Clifford P., and J. David Gopoian. "Dispositional empathy and the political gender gap." *Women & Politics* 21, no. 2 (2000): 1–20.

Phua, Desiree Y., Helen Chen, Fabian Yap, Yap Seng Chong, Peter D. Gluckman, Birit F. P. Broekman, Johan G. Eriksson, and Michael J. Meaney. "Allostatic load in children: The cost of empathic concern." *Proceedings of the National Academy of Sciences* 120, no. 39 (2023): e2217769120.

Ritter, Ryan S., Jesse Lee Preston, and Ivan Hernandez. "Happy tweets: Christians are happier, more socially connected, and less analytical than atheists on Twitter." *Social Psychological and Personality Science* 5, no. 2 (2014): 243–249.

Skitka, Linda J., and Philip E. Tetlock. "Allocating scarce resources: A contingency model of distributive justice." *Journal of Experimental Social Psychology* 28, no. 6 (1992): 491–522.

Tetlock, Philip E. "Social functionalist frameworks for judgment and choice: Intuitive politicians, theologians, and prosecutors." *Psychological Review* 109, no. 3 (2002): 451.

Tetlock, P. E., O. V. Kristel, S. B. Elson, M. C. Green, J. S. Lerner. "The psychology of the unthinkable: Taboo trade-offs, forbidden base rates, and heretical counterfactuals." *Journal of Personality and Social Psychology* 78, no. 5 (2000): 853–870.

CHAPTER 8:

Barrick, Murray R., and Michael K. Mount. "The big five personality dimensions and job performance: A meta-analysis." *Personnel Psychology* 44, no. 1 (1991): 1–26.

Davis, Leslie, and Kim Parker. "A half-century after 'Mister Rogers' debut, 5 facts about neighbors in U.S." Pew Research Center, August 15, 2019. https://www.pewresearch.org/fact-tank/2019/08/15/facts-about-neighbors-in-u-s/.

Haerpfer, C., R. Inglehart, A. Moreno, C. Welzel, K. Kizilova, J. Diez-Medrano, M. Lagos, P. Norris, E. Ponarin, and B. Puranen. "World Values Survey Wave 7" (2017–2022) Cross-National Data-Set. Version: 4.0.0. World Values Survey Association, 2022. DOI: doi.org/10.14281/18241.18.

Henrich, Joseph. *The WEIRDest People in the World: How the West Became Psychologically Peculiar and Particularly Prosperous*. London: Penguin, 2020.

Jennings, Justin. *Globalizations and the Ancient World*. Cambridge, UK: Cambridge University Press, 2010.

Jung, Haesung, and Dolores Albarracín. "Concerns for others increase the likelihood of vaccination against influenza and COVID-19 more in sparsely rather than densely populated areas." *Proceedings of the National Academy of Sciences* 118, no. 1 (2021): e2007538118.

Nissen, H. J. "The archaic texts from Uruk." *World Archaeology* 17 (1986): 317–334.

Parker, Kim, Juliana Menasce Horowitz, Anna Brown, Richard Fry, D'Vera Cohn, and Ruth Igielnik. "6. How urban, suburban and rural residents interact with their neighbors." Pew Research Center, May 22, 2018. https://www.pewresearch.org/social-trends/2018/05/22/how-urban-suburban-and-rural-residents-interact-with-their-neighbors/.

Ritchie, Hannah. "How urban is the world?" Our World in Data, September 17, 2018. https://ourworldindata.org/how-urban-is-the-world#:~:text=Using%20these%20definitions%2C%20it%20reports,shown%20in%20the%20chart%20here.

Ritchie, Hannah, Veronika Samborska, and Max Roser. "Urbanization." Our World in Data, February 2024. https://ourworldindata.org/urbanization#urbanization-over-the-past-500-years.

CHAPTER 9:

Berkessel, Jana B., Jochen E. Gebauer, Mohsen Joshanloo, Wiebke Bleidorn, Peter J. Rentfrow, Jeff Potter, and Samuel D. Gosling. "National religiosity eases the psychological burden of poverty." *Proceedings of the National Academy of Sciences* 118, no. 39 (2021): e2103913118.

Borman, Geoffrey D., Christopher S. Rozek, Jaymes Pyne, and Paul Hanselman. "Reappraising academic and social adversity improves middle school students' academic achievement, behavior, and well-being." *Proceedings of the National Academy of Sciences* 116, no. 33 (2019): 16286–16291.

Chetty, Raj, John N. Friedman, Nathaniel Hilger, Emmanuel Saez, Diane Whitmore Schanzenbach, and Danny Yagan. "How does your kindergarten classroom affect your earnings? Evidence from Project STAR." *Quarterly Journal of Economics* 126, no. 4 (2011): 1593–1660.

Darwin, C. *On the Origin of Species*. London: John Murray, 1859.

Dietze, Pia, and Eric D. Knowles. "Social class predicts emotion perception and perspective-taking performance in adults." *Personality and Social Psychology Bulletin* 47, no. 1 (2021): 42–56.

Field, Erica, Vera Molitor, Alice Schoonbroodt, and Michèle Tertilt. "Gender gaps in completed fertility." *Journal of Demographic Economics* 82, no. 2 (2016): 167–206.

General Social Survey, NORC, University of Chicago.

Harati, Hamidreza, and Thomas Talhelm. "Cultures in water-scarce environments are more long-term oriented." *Psychological Science* 34, no. 7 (2023): 754–770.

Iyengar, Sheena S., and Mark R. Lepper. "When choice is demotivating: Can one desire too much of a good thing?" *Journal of Personality and Social Psychology* 79, no. 6 (2000): 995.

Kebede, Endale, Anne Goujon, and Wolfgang Lutz. "Stalls in Africa's fertility decline partly result from disruptions in female education." *Proceedings of the National Academy of Sciences* 116, no. 8 (2019): 2891–2896.

Kern, Margaret L., Paul X. McCarthy, Deepanjan Chakrabarty, and Marian-Andrei Rizoiu. "Social media-predicted personality traits and values can help match people to their ideal jobs." *Proceedings of the National Academy of Sciences* 116, no. 52 (2019): 26459–26464.

Knight, Caroline, Doina Olaru, Julie Lee, and Sharon Parker. "The loneliness of the hybrid worker." *MIT Sloan Management Review* 63, no. 4 (2022): 10–12.

Kraus, Michael W., and Dacher Keltner. "Signs of socioeconomic status: A thin-slicing approach." *Psychological Science* 20, no. 1 (2009): 99–106.

Piff, Paul K., Daniel M. Stancato, Andres G. Martinez, Michael W. Kraus, and Dacher Keltner. "Class, chaos, and the construction of community." *Journal of Personality and Social Psychology* 103, no. 6 (2012): 949.

Ross, Catherine E., and Chia-ling Wu. "The links between education and health." *American Sociological Review* 60, no. 5 (1995): 719–745.

CHAPTER 10:

Andersson, Gunnar. "Trends in marriage formation in Sweden 1971–1993." *European Journal of Population/Revue européenne de Démographie* 14 (1998): 157–178.

Apostolou, Menelaos. "Sexual selection under parental choice: The role of parents in the evolution of human mating." *Evolution and Human Behavior* 28, no. 6 (2007): 403–409.

Ayers, David J. "The Gender Gap in Marriages Between College-Educated Partners." Institute for Family Studies, October 23, 2019. https://ifstudies .org/blog/the-gender-gap-in-marriages-between-college-educated-partners.

Butterworth, J., S. Pearson, and W. von Hippel. "Dual mating strategies observed in male clients of female sex workers." *Human Nature* 34, no.1 (2023): 46–63.

Finkel, Eli J., Chin Ming Hui, Kathleen L. Carswell, and Grace M. Larson. "The suffocation of marriage: Climbing Mount Maslow without enough oxygen." *Psychological Inquiry* 25, no. 1 (2014): 1–41.

Jia, Haomiao, and Erica I. Lubetkin. "Life expectancy and active life expectancy by marital status among older US adults: Results from the US Medicare Health Outcome Survey (HOS)." *SSM-Population Health* 12 (2020): 100642.

Joel, Samantha, Paul W. Eastwick, Colleen J. Allison, Ximena B. Arriaga, Zachary G. Baker, Eran Bar-Kalifa, Sophie Bergeron et al. "Machine learning uncovers the most robust self-report predictors of relationship quality across 43 longitudinal couples studies." *Proceedings of the National Academy of Sciences* 117, no. 32 (2020): 19061–19071.

Joel, Samantha, Paul W. Eastwick, and Eli J. Finkel. "Is romantic desire predictable? Machine learning applied to initial romantic attraction." *Psychological Science* 28, no. 10 (2017): 1478–1489.

Kiecolt-Glaser, Janice K., and Tamara L. Newton. "Marriage and health: His and hers." *Psychological Bulletin* 127, no. 4 (2001): 472.

Kramer, Karen L., Ryan Schacht, and Adrian Bell. "Adult sex ratios and partner scarcity among hunter–gatherers: implications for dispersal patterns and the evolution of human sociality." *Philosophical Transactions of the Royal Society B: Biological Sciences* 372, no. 1729 (2017): 20160316.

Lichter, Daniel T., Joseph P. Price, and Jeffrey M. Swigert. "Mismatches in the marriage market." *Journal of Marriage and Family* 82, no. 2 (2020): 796–809.

Lucas, Richard E. "Time does not heal all wounds: A longitudinal study of reaction and adaptation to divorce." *Psychological Science* 16, no. 12 (2005): 945–950.

Lucas, Richard E., Andrew E. Clark, Yannis Georgellis, and Ed Diener. "Reexamining adaptation and the set point model of happiness: reactions to changes in marital status." *Journal of Personality and Social Psychology* 84, no. 3 (2003): 527.

Marlowe, Frank W. "Mate preferences among Hadza hunter-gatherers." *Human Nature* 15, no. 4 (2004): 365–376.

https://www.pewresearch.org/social-trends/2020/08/20/a-profile-of-single -americans/.

Robles, Theodore F., Richard B. Slatcher, Joseph M. Trombello, and Meghan M. McGinn. "Marital quality and health: a meta-analytic review." *Psychological Bulletin* 140, no. 1 (2014): 140.

CHAPTER 11:

Atalay, Enghin. "A twenty-first century of solitude? Time alone and together in the United States." (2022). Working Paper, Federal Reserve Bank of Philadelphia.

Bargh, John A., and Katelyn YA McKenna. "The Internet and social life." *Annual Review of Psychology* 55 (2004): 573–590.

Hasson, Uri, and Chris D. Frith. "Mirroring and beyond: coupled dynamics as a generalized framework for modelling social interactions." *Philosophical Transactions of the Royal Society B: Biological Sciences* 371, no. 1693 (2016): 20150366.

Kang, Olivia, and Thalia Wheatley. "Pupil dilation patterns spontaneously synchronize across individuals during shared attention." *Journal of Experimental Psychology: General* 146, no. 4 (2017): 569.

Kendon, Adam. "Some functions of gaze-direction in social interaction." *Acta Psychologica* 26 (1967): 22–63.

Kiesler, S., J. Siegel, and T. McGuire. "Social psychological aspects of computer-mediated communication." *American Psychologist* 39, no. 10 (1984): 1129–1134.

Koul, Atesh, Davide Ahmar, Gian Domenico Iannetti, and Giacomo Novembre. "Spontaneous dyadic behaviour predicts the emergence of interpersonal neural synchrony." *NeuroImage* 277 (2023): 120233.

Lakin, Jessica L., and Tanya L. Chartrand. "Using nonconscious behavioral mimicry to create affiliation and rapport." *Psychological Science* 14, no. 4 (2003): 334–339.

Leong, Victoria, Elizabeth Byrne, Kaili Clackson, Stanimira Georgieva, Sarah Lam, and Sam Wass. "Speaker gaze increases information coupling between infant and adult brains." *Proceedings of the National Academy of Sciences* 114, no. 50 (2017): 13290–13295.

Levine, Hagai, Niels Jørgensen, Anderson Martino-Andrade, Jaime Mendiola, Dan Weksler-Derri, Maya Jolles, Rachel Pinotti, and Shanna H. Swan. "Temporal trends in sperm count: a systematic review and meta-regression analysis of samples collected globally in the 20th and 21st centuries." *Human Reproduction Update* 29, no. 2 (2023): 157–176.

McKenna, Katelyn Y. A., and J. A. Bargh. "Coming out in the age of the

Internet: Identity 'demarginalization' through virtual group participation." *Journal of Personality and Social Psychology* 75, no. 3 (1998): 681–694.

McKenna, Katelyn Y. A., A. S. Green, and M. J. Gleason. "Relationship formation on the Internet: What's the big attraction?," *Journal of Social Issues* 58, no. 1 (2002): 9–31.

Newman, David B., John B. Nezlek, and Todd M. Thrash. "The dynamics of searching for meaning and presence of meaning in daily life." *Journal of Personality* 86, no. 3 (2018): 368–379.

Oberman, Lindsay M., Piotr Winkielman, and Vilayanur S. Ramachandran. "Face to face: Blocking facial mimicry can selectively impair recognition of emotional expressions." *Social Neuroscience* 2, no. 3–4 (2007): 167–178.

Pryluk, Raviv, Yosef Shohat, Anna Morozov, Dafna Friedman, Aryeh H. Taub, and Rony Paz. "Shared yet dissociable neural codes across eye gaze, valence and expectation." *Nature* 586, no. 7827 (2020): 95–100.

Rosa, Hartmut. *Resonance: A Sociology of Our Relationship to the World.* Hoboken, NJ: John Wiley & Sons, 2019.

Templeton, Emma M., Luke J. Chang, Elizabeth A. Reynolds, Marie D. Cone LeBeaumont, and Thalia Wheatley. "Fast response times signal social connection in conversation." *Proceedings of the National Academy of Sciences* 119, no. 4 (2022): e2116915119.

Ueda, Peter, Catherine H. Mercer, Cyrus Ghaznavi, and Debby Herbenick. "Trends in frequency of sexual activity and number of sexual partners among adults aged 18 to 44 years in the US, 2000-2018." *JAMA Network Open* 3, no. 6 (2020): e203833.

Ulmer Yaniv, Adi, Roy Salomon, Shani Waidergoren, Ortal Shimon-Raz, Amir Djalovski, and Ruth Feldman. "Synchronous caregiving from birth to adulthood tunes humans' social brain." *Proceedings of the National Academy of Sciences* 118, no. 14 (2021): e2012900118.

van Baar, Jeroen M., David J. Halpern, and Oriel FeldmanHall. "Intolerance of uncertainty modulates brain-to-brain synchrony during politically polarized perception." *Proceedings of the National Academy of Sciences* 118, no. 20 (2021): e2022491118.

Wohltjen, Sophie, and Thalia Wheatley. "Eye contact marks the rise and fall of shared attention in conversation." *Proceedings of the National Academy of Sciences* 118, no. 37 (2021): e2106645118.

Zhao, Nan, Xian Zhang, J. Adam Noah, Mark Tiede, and Joy Hirsch. "Separable processes for live 'in-person' and live 'Zoom-like' faces." *Imaging Neuroscience* 1 (2023): 1–17.

CHAPTER 12:

Buyalskaya, Anastasia, Hung Ho, Katherine L. Milkman, Xiaomin Li, Angela L. Duckworth, and Colin Camerer. "What can machine learning teach us about habit formation? Evidence from exercise and hygiene." *Proceedings of the National Academy of Sciences* 120, no. 17 (2023): e2216115120.

Gollwitzer, Peter M. "Implementation intentions: strong effects of simple plans." *American Psychologist* 54, no. 7 (1999): 493.

Khera, Amit V., Mark Chaffin, Kaitlin H. Wade, Sohail Zahid, Joseph Brancale, Rui Xia, Marina Distefano et al. "Polygenic prediction of weight and obesity trajectories from birth to adulthood." *Cell* 177, no. 3 (2019): 587–596.

Murthy, Vivek H. *Together: Loneliness, Health and What Happens When We Find Connection.* New York: HarperCollins, 2020.

Verplanken, Bas, and Sheina Orbell. "Attitudes, habits, and behavior change." *Annual Review of Psychology* 73 (2022): 327–352.

Index

Entries in *italics* refer to figures and illustrations.

About the Author

WILLIAM VON HIPPEL grew up in Alaska, went to Yale and the University of Michigan, and then taught psychology for a dozen years at Ohio State University before finding his way to Australia, where he taught at the University of New South Wales and the University of Queensland for twenty-one years. He has published over 150 research articles, chapters, and edited books, and is the author of *The Social Leap*. His work has been featured in media outlets such as the *New York Times*, the BBC, *USA Today*, the *Economist*, *Le Monde*, *Il Mondo*, *Der Spiegel*, and the *Australian*. He lives in Brisbane, Australia.